Sources and Studies in the History of Mathematics and Physical Sciences

More information about this series at http://www.springer.com/series/4142

Sources and Studies in the History of Mathematics and Physical Sciences was inaugurated as two series in 1975 with the publication in Studies of Otto Neugebauer's seminal three-volume History of Ancient Mathematical Astronomy, which remains the central history of the subject. This publication was followed the next year in Sources by Gerald Toomer's transcription, translation (from the Arabic), and commentary of Diocles on Burning Mirrors. The two series were eventually amalgamated under a single editorial board led originally by Martin Klein (d. 2009) and Gerald Toomer, respectively two of the foremost historians of modern and ancient physical science. The goal of the joint series, as of its two predecessors, is to publish probing histories and thorough editions of technical developments in mathematics and physics, broadly construed. Its scope covers all relevant work from pre-classical antiquity through the last century, ranging from Babylonian mathematics to the scientific correspondence of H. A. Lorentz. Books in this series will interest scholars in the history of mathematics and physics, mathematicians, physicists, engineers, and anyone who seeks to understand the historical underpinnings of the modern physical sciences.

Carlos Gonçalves

Mathematical Tablets from Tell Harmal

Springer

Carlos Gonçalves
School of Arts, Sciences and Humanities
University of São Paulo
São Paulo, São Paulo, Brazil

ISSN 2196-8810 ISSN 2196-8829 (electronic)
Sources and Studies in the History of Mathematics and Physical Sciences
ISBN 978-3-319-37404-8 ISBN 978-3-319-22524-1 (eBook)
DOI 10.1007/978-3-319-22524-1

Mathematics Subject Classification (2010): 01A17

Springer Cham Heidelberg New York Dordrecht London

Softcover re-print of the Hardcover 1st edition 2015

Printed on acid-free paper

Springer International Publishing AG Switzerland is part of Springer Science+Business Media (www.springer.com)

To Estela

Acknowledgements

During the production of this book, I received direct support from many researchers in history of mathematics and Assyriology. I am thankful to all of them for their effort to help me trying to make this a good and interesting text. In spite of that, the reader may still find here things to be improved or amended. These are, of course, my own responsibility.

I thank Hermann Hunger for kindly receiving me during the second semester of 2009, as a visitor to the Institut für Orientalistik of the University of Vienna.[1] This was a period of great importance to improving my understanding of the language of the mathematical texts. I am also thankful for his reading of a preliminary version of this work and for his criticisms. During my stay at the IfO, some other people also helped me in various ways. In particular, I am indebted to Michael Jursa and Michaela Weszeli.

I also profited from a rich exchanging of ideas with Jens Høyrup, who diligently and critically read the book. I am deeply indebted to him for his suggestions and amendments, as well as for his enthusiasm and availability. I am especially thankful for his visit to the University of São Paulo, during the month of June of 2013.[2]

In Paris, I received much help from Christine Proust, namely, countless pieces of advice on the sexagesimal and the measure system that saved me from drowning in cuneiform numbers. Christine Proust, Agathe Keller and Karine Chemla made me feel welcome in the Sciences in the Ancient World (SAW) project, carried out in the SPHERE Laboratory (Paris 7, CNRS). I have been fortunate enough to attend several meetings in the SAW-SPHERE context, where many people offered me some contribution that was absorbed in this book: I thank Camille Lecompte, Grégoire Nicolet, Laurent Colonna d'Istria, Robert Midekke-Conlin and Xiaoli Ouyang for that.[3]

[1] With the financial support of the São Paulo state funding agency FAPESP (Fundação de Amparo à Pesquisa do Estado de São Paulo).

[2] Thanks again to FAPESP.

[3] I thank the *Mairie de Paris* for a scholarship under the program Research in Paris, that allowed me to stay in the city for some months in 2013.

Still in Paris, I could profit from the active local Assyriological community. Above all, I thank Cécile Michel for her support.

Much important too was the help from people that I contacted through email or that I approached during congresses and seminars: Antoine Cavigneaux, Duncan Melville, Eleanor Robson, Karen Nemet-Nejat and Jöran Friberg.

In Brazil, colleagues and friends from very different areas supported my work, sharing with me their views about history and studies of culture: Gérard Grimberg, Gert Schubring, Fumikazu Saito, Jefferson Mello, José Luiz Goldfarb, Márcia de Barros Silva, Maria Beatriz Florenzano, Maria Isabel Fleming, Pablo Ortellado, Patrícia Valim, Thomás Haddad, Rogério de Siqueira, Ulpiano de Meneses and Vera Machline. I am particularly indebted to the Assyriological exchange with my friend Marcelo Rede.

Furthermore, I am grateful for the precious support I had from Selma Mascagna, Estela's mother. I am also thankful to my brother João Gonçalves, a Sanskrit expert who has always been available to hear cuneiform stories.

Last but not least, I thank Tatiana Roque for helping me in textual and philosophical ways and, above all, for being who she is in my life.

Contents

Introduction

The purpose of this work is to present together a group of 12 mathematical tablets from the site of Tell Harmal, the modern denomination of the ancient city of Šaduppûm. Originally, these tablets were published by Taha Baqir in *Sumer*, the Iraqi journal of archaeology, in the beginning of the 1950s. The titles of the papers written by Baqir, "An Important Mathematical Problem Text from Tell Harmal" (Baqir 1950a), "Another Important Mathematical Text from Tell Harmal" (Baqir 1950b) and "Some More Mathematical Texts from Tell Harmal" (Baqir 1951), inspire the title of the present exposition, *Mathematical Tablets from Tell Harmal.*

Besides making these tablets available together for the first time, this book also brings another contribution, about which I should write a few words.

The point of departure is that since the original edition of these tablets in the early 1950s, several new studies on Mesopotamian mathematics have been published. These new studies have added to our knowledge of the cuneiform mathematical practices: on the one hand, they have established a new approach to the study of mathematical terminology and procedures; on the other hand, they have started to draw more delineated links between the mathematical practices and the larger society. It could not go without saying that, as a matter of fact, a large number of new studies and approaches to Assyriology in general have come to light in the last 50 years and that part of these new scholar works has its impact on the history of Mesopotamian mathematics. However, it is not my intention here to write a history of the field, and that is why I refrain from quoting specific works that would have been most influential in shaping the discipline as it is today. In spite of this reserve of mine, two key contributions to the history of Mesopotamian mathematics must be nominally acknowledged, that of the geometric interpretation, in terms of cut-and-paste geometry, of the so-called "Babylonian algebra" (see Chap. 2) and that of conformal translation (see Chap. 3) as a productive approach to the cuneiform mathematical texts. It is no exaggeration to say that without these two approaches, the present work would perhaps be of much lesser interest.

Thus, my choice of the tablets to be dealt with here was based mainly on their potential to feed a study that would not be the mere repetition of what one can already find in the original publications. With this end in view, this equaled to studying tablets that had been published accompanied by autograph copies and by photographs, so that I could be sure of applying the new views of the history of mathematics to texts I could check with a reasonable degree of safety. As a side effect, I was enabled to venture a very few new readings which I considered relevant to include in a work so language-oriented as the present one.

The non-availability of photographs prevented me from including here the very important so-called mathematical compendium (Goetze 1951), also from Tell Harmal. As the tablets of the compendium—indeed perhaps two very similar compendia—are in an extremely lacunar state, and the publication by Goetze makes this clear, it would have been a daring experiment to try to improve the reading. Besides, a work in this direction has already been conducted (Robson 1999, 196–199).

As for the mathematical tablets published by Bruins in *Sumer* (1953b, 1954), it must be noticed initially that not all of them come from Tell Harmal. Furthermore, they were published without a proper visual apparatus, that is, without either copies or photographs. Thus, these texts have not been included here. Bruins, however, also published articles in which he suggested, sometimes very harshly, improvements and corrections pertaining to the 12 tablets I study in this work. I have considered fit to take into account these papers, although I was not always able to agree with what they conveyed.

The present work contains six chapters of varying length that can be read in different orders by different readers.

In Chap. 1, I deal briefly with the site of Tell Harmal and the dating of the tablets. Besides informing the readers about the chronological issue, the chapter has the goal of keeping us all aware that mathematical texts are indeed mathematical clay tablets, material objects coming from a modern archaeological site which corresponds to an ancient place. This very short chapter is surely one possible starting point for the reader.

Chapter 2 is devoted to summarising the mathematical contents found in the dozen tablets examined in the work. It is true that each tablet can be read independently and that I have made, in Chap. 4, every effort to write the individual mathematical commentaries as self-contained texts. However, it is also true that there are common points between these tablets. Chapter 2 is a discussion of these points.

Chapter 3 contains the conventions for the philological treatment of the tablets and an itemised exposition of the language problems encountered in the analysed corpus. Some of these problems have already been fervently discussed by other historians of Mesopotamian mathematics. In these cases, I tried to give the reader a number of useful references. The remaining problems discussed in Chap. 3 issued from the specific options of this work.

Chapters 2 and 3 are conceived as prerequisites for understanding the main material, that is, the 12 mathematical tablets from Tell Harmal that I analyse here. The reader might of course just scan these chapters quickly, skipping the sections or the paragraphs that do not offer interest in a first reading. Then, while examining the mathematical tablets in Chap. 4, one can come back to Chaps. 2 and 3 to see in detail the mathematical and the language issues involved or to check how conventions are used.

Chapter 4 represents the most important part of the book. In it, the 12 selected tablets are presented and analysed. Each tablet receives initially a transliteration and a transcription:

- A transliteration of the cuneiform signs identifies the signs by their conventional names and tries to respect their isolated pronunciation.
- A transcription renders morphological and syntactical aspects of the original text (see also Chap. 3 for the details and conventions governing each of these parts).[1]

The tablets are then translated and commented. The only exception is IM54010, whose bad state of conservation prevents a complete understanding.

Transliteration and transcription are directed to specialists in cuneiform mathematics or to readers who possess at least a basic command of the Akkadian and a desire to improve their understanding of the original texts. Translation and commentary, on the other hand, were designed to satisfy the readers' appetite for the mathematical ideas and techniques that these texts bear.

Besides, in this chapter, I also offer some philological remarks and a mathematical analysis for each tablet.

Chapter 5 describes what I believe to be some of the central problems of the historiography of mathematics in general, but presented here almost exclusively in relation to the Old Babylonian mathematics. It discusses the ways history of mathematics has typically dealt with the mathematical evidence and inquires how and to what degree mathematical tablets can be made part of a picture of the larger social context; furthermore, it gives a contribution to a geography of the Old Babylonian mathematical practices, by evidencing that scribes in Šaduppûm made use of cultural material that was locally available. This is an almost independent chapter. It may as well be a starting point, if the reader is interested in getting a general image of Old Babylonian mathematics in its sociocultural context.

Finally, Chap. 6 is the vocabulary of the whole textual corpus. It is a chapter to be used as a reference, if one wants to see where a given word is used or to know the semantic field to which it belongs in the general language.

All in all, this book is a rereading of a group of 4000-year-old mathematical tablets published in the 1950s from the point of view of the historiography developed from the 1990s on. It is thus an application of the new lessons of the history of mathematics. I tried to keep the language issues as friendly as possible to the non-initiated, but I am aware that they require from the reader some will in order to be overcome. Were I to give a piece of advice, I would suggest that the approximation to the language questions be made gradually. The difficulties will fade out as the reader advances. One should keep calm, if I may say this, and the experience of visiting a different mathematical culture in its own terms will be a most rewarding one.

[1] This is a practice, however, not usual in works like the present one. This is due mainly to the fact that, in general, experienced readers of Akkadian are able to produce mentally the transcriptions when they need so. Nevertheless, as this work may also be read by intermediate students of Akkadian, I considered it fit to leave the transcriptions here. Knowing—much by experience—that transcriptions are very prone to error, I made great effort to establish them as correct as possible. Notices about remaining mistakes or inaccuracies will be happily welcome.

Chapter 1
The Site of Tell Harmal and the Archaeological Record

The site of Tell Harmal, which corresponds to the ancient city of Šaduppûm, is located in the vicinities of Baghdad, near Tell Mohammed and in the angle between the Tigris and the Diyala. It has around 150 m of diameter, and by the time of the first excavations it rose about 4 m above the level of the surrounding ground.

Excavations were conducted under the responsibility of the Directorate-General of Antiquities of Iraq during the following periods: 1945–1949, 1958–1959 and 1962–1963 (Baqir 1959, 3; 1961, 4; Miglus 2007, 2008, 491). A pair of further campaigns, carried out in 1997 and 1998, was the result of a cooperation between the University of Baghdad and the German Archaeological Institute (Hussein and Miglus 1998, 1999). The tablets studied in the present work were exhumed during the first phase of excavations, that is to say, in the period from 1945 to 1949.

The stratigraphy of the site was first described in the section "Notes & Statistics" of the issue of January 1946 of *Sumer* (1946, vol. II). A more detailed exposition was published soon after by Baqir (1946). The stratigraphy was once more mentioned in a two-page note (Baqir 1948) and a little more thoroughly discussed and expanded in two papers dealing with the date formulas found in the site (Baqir 1949a, b). A booklet about Tell Harmal was published in Arabic and in English by the Directorate-General of Antiquities of Iraq, condensing all the results (Baqir 1959). As an outcome of the excavations carried out jointly by the University of Baghdad and the German Archaeological Institute, new questions and new findings about the stratigraphy of the site were published (Hussein and Miglus 1998, 1999). All the previous results since the 1940s are organised and described in the article "Šaduppûm B" of the *Reallexikon der Assyriologie* (Miglus 2007, 2008).

The seven occupational levels of the site are divided as follows (Miglus 2007, 2008):

> Levels VII and VI (possibly, in part, also Level V) date to the 3rd millennium, Levels V-II date to the first quarter of the 2nd millennium (Isin-Larsa Period) and Level I dates to the second half of the 2nd millennium (Kassite Period)

C. Gonçalves, *Mathematical Tablets from Tell Harmal*, Sources and Studies in the History of Mathematics and Physical Sciences,
DOI 10.1007/978-3-319-22524-1_1

The excavations showed that in the time of levels III and II (within the Isin-Larsa Period), the city was surrounded by a 5-m thick wall with the shape of a trapezium. Within the enclosed area, there was a large temple, a few shrines, private houses, a large administrative building and what seemed to be a line of offices of professional scribes (Baqir 1959). The mathematical tablets I present in this work all come from this context and period.

Baqir (1959) reports that, according to the date formulas found in the site, Level II dates to the time of Daduša and Ibal-piʾel II, kings of Ešnunna, the latter being a contemporary of Hammurabi, while Level III dates to the time of Ipiq-Adad II, also from Ešnunna.

To date, no complete report on the excavations at Tell Harmal has ever been published. As a consequence, it is not possible to know the exact places where the mathematical tablets I analyse in this work were found. This means that, although the numbers of the rooms where these tablets come from are known, it is not possible to locate physically their provenances on the published partial maps of the site (Baqir 1946, 1959; Miglus 2007, 2008).

Laith Hussein (2009) studied the texts from the so-called "Serai", the large administrative building of Šaduppûm. Speaking about the texts from that ancient city in general, the author reported that until that moment there were around 600 texts that had been published. A list gives for each publication the number of the rooms where the corresponding tablets were found, as well as the stratigraphic levels to which they belong. As regards the mathematical texts, there is in this list one inconsistency with the same data presented by Baqir in his publications: tablet IM54559 is reported to come from room 256 by Baqir and from room 252 by Hussein. Table 1.1 summarises the places and the stratigraphic levels where the tablets were found. It also specifies the mentioned inconsistency.

Table 1.1 The analysed tablets and their places of exhumation

Tablet	Year of discovery	Room (according to Baqir)	Room (according to Hussein 2009)	Stratigraphic level (according to Baqir)
IM55357	1949[a]	Not given	301	Level III
IM52301	Not given[b]	180	180	Level II
IM54478	The fourth season of work, 1949 (Baqir 1951)	Nine texts come from room 252	All these texts come from room 252	10 cm beneath the pavement of Level II
IM53953				
IM54538				
IM53961				
IM53957				
IM54010				
IM53965				
IM54464				
IM54011				
IM54559		256		

[a]Baqir (1950a, 39) states that the tablet was found "during this season's dig", by which I understand the 1949 excavation. In another place, Baqir (1949b, 136) describes the "fifth season" as having lasted from September, 1949 until the end of November of the same year

[b]Baqir (1950b, 130) states that the tablet was found "during the dig of one of the previous seasons"

In Chap. 4, I give the approximate measures of these tablets. Finally, it is worth mentioning that IM55357 and IM52301 use portrait orientation, while the other ten are landscape. Whether this can be related to their possible pedagogical function is something still to be understood (Gonçalves 2010). The only piece of information available at the moment is that room 252 is most likely situated in a private residence (Baqir 1951), a fact that by itself proves nothing, but is consistent with the observation that, during the Old Babylonian period, some places of scribal education in other cities were also private residences.

Chapter 2
A Few Remarks About Old Babylonian Mathematics

This chapter contains material dealing with a number of technical aspects of Old Babylonian mathematics, ranging from the vocabulary used by scribes to the techniques they employed to solve problems. My intention is to give the reader some of the fundamental tools to interpret the mathematics of the tablets according to the present knowledge about Old Babylonian mathematics. This entails that we should pay attention, for instance, to the different meanings and modes of expression that each numerical operation had. Or to the fact that the mathematical techniques employed in the past free us, in a certain measure, from the temptation of using our symbolic school algebra to interpret the contents of the tablets. All in all, the discussion of these features is necessary in order to characterise the cuneiform mathematical tradition, as much as possible, in its own terms.

2.1 Words for Arithmetical Operations

This section describes the vocabulary of arithmetical operations. The reader not yet used to Old Babylonian mathematics should pay special attention to the following paragraphs, as they assume that arithmetic operations were conceptualised quite differently from what is done in our own arithmetics. For the sake of convenience, I list only words that are present in the analysed tablets. For other common arithmetical vocabulary, the reader may refer to Høyrup (2002, 2010).

Additive operations are indicated by the Akkadian verbs *waṣābum* (logogramically taḫ$_2$), add, and *kamārum*, accumulate. According to Høyrup (2002, 19; 2010, 399), the first one indicates an absorption of one thing into another, being thus an asymmetric operation and dealing with two addends that are thought of as having the same nature. The second operation, on the other hand, is a symmetric one and tends to be used when things added are considered only in their numerical values, either having different natures or not.

C. Gonçalves, *Mathematical Tablets from Tell Harmal*, Sources and Studies in the History of Mathematics and Physical Sciences,
DOI 10.1007/978-3-319-22524-1_2

The tablets analysed here are in general consistent with the above picture. During the process of solving problems about squares (the procedure is explained below in Sect. 2.2), a square is added (*waṣābum,* taḫ$_2$) to a certain geometrical figure, producing a new square, an operation that may be interpreted as an absorption of the smaller region into the other one. Also, in the final steps of some problems, where two numerical answers are obtained by adding and cutting one number to and off another, the absorption metaphor applies. The same occurs in IM54464, where the scribe adds a small quantity of silver to a larger one, conserving, so to say, the identity of the bigger amount.

Consistently, accumulation is used to deal with things in a more symmetric way, as in the following passages: you accumulate one cubit and one broken cubit (IM54011); accumulate 0;4 and 0;4, and 0;8 comes up (IM54464). The intention in these cases seems to consider only the formal numerical results of the accumulations.

However, the distinction between these two verbs need not be taken as a rigid one, reflecting some ultimate nature of the involved quantities. For instance, in the statement of Problem 2 of tablet IM52301, we read "If I added 10 [...] to the two thirds of the accumulation of the upper and lower width, I built" a second length. This addition of linear measures is, in a certain sense, the same situation of tablet IM54011 mentioned in the last paragraph and denotes either a chronological difference in terminology (IM52301 is older than IM54011) or the possibility a scribe had of seeing an operation as accumulation or addition according to what he[1] wanted to convey: in this respect, it is remarkable that in IM52301, a new length is built, which indicates that the operation was interpreted as an absorption of one, so to say, linear measure into another, whereas in IM54011 the scribe seems to have been interested only in the numerical value of the operation.

In the analysed tablets, the word for the result of the verb *kamārum* is *kumurrûm*, accumulation. Not associated with any of the two additive verbs, the vocabulary also includes the less conventional *napḫarum*, sum, occurring in the unconventional Problem 3 of IM52301.

The semantic field of subtraction is covered by *ḫarāṣum*, cut off, and *nasāḫum*, remove. They are, by the way, used interchangeably in IM54464. There is, in IM55357, the composite logogram ib$_2$.tag$_4$.a, associated with the Akkadian verbs *ezēbum*, remain, and *riāḫum*, be left behind, thus indicating the result of a subtraction. The result is further indicated by *šapiltum* and *šittum*, both translated as remainder.

The main multiplicative operation is expressed by the verb *našûm*, raise. For instance, in IM55357, the scribe raises 0;45 to 2 and obtains 1;30.[2]

[1] Shamefully, the scribe is always referred to as "he" in this work. There seems to be no indication of women scribes in Šhaduppûm, and scribal activity in Mesopotamia was, in general, restricted to men. The women's apartments in the royal palace of Mari and a group of women scribes from Sippar are exceptions to this rule (Lion and Robson 2005).

[2] Numbers are written in base 60. I will often use the comma "," as digit separator and the semicolon ";" as the separator of the integral and fractional parts. See in what follows more details about this as well as a different approach in the Sects. 2.5 and 2.6.

Another multiplication is given by a form of *akālum*, eat: the form is its Št-Stem *šutākulum*, combine.[3] For example, in IM53957, the scribe combines two thirds and two thirds and obtains 0;26,40. Notice that the last operation is equivalent to obtaining the square of a number.

Finally, in the statement of IM54478, we see a form of the verb *maḫārum*—accept, approach, face, rival, be equal: the form is its lexical Št-Stem *šutamḫurum*, cause things to confront each other, an expression that is used to designate the construction of a square (Høyrup 2002, 25).

Square and cube roots are referred to with the Sumerian terms $ib_2.si_8$ (maybe to be read $ib_2.sa_2$), its variation ib_2.si.e and ba.si.e, all bringing the idea of "being/making equal", when used as verbs, or "the equal", when used as nouns.[4] The last term also occurs in an Akkadianised form, *basûm*, the equal. For instance, in IM52301 (line 9), the scribe causes to come up the equal of 2,46,40, and obtains 1,40. These terms are further discussed in Sect. 3.2 from a more linguistic point of view.

In order to break a quantity in its half, scribes use *ḫepûm*, halve. The half so produced is called *bāmtum*, which occurs in IM55357.[5] Another term to designate half of something is *zūzum*. On the other hand, doubling is expressed with the verb *eṣēpum*, double.

The last operation occurring in the tablets analysed in this work is that of taking a reciprocal of a number. The verb for this is *paṭārum*, detach. For instance, in IM52301, the scribe detaches the *igi* of 0;45 and obtains 1;20. The reciprocal itself is called *igi* and is left untranslated in this work, for reasons explained in Chap. 3.

2.2 Cut-and-Paste Geometry and Problems About Squares

In *Lengths, Widths and Surfaces*, Jens Høyrup (2002) used the expression "cut-and-paste geometry" to describe a technical aspect of Old Babylonian mathematics. In fact, cut-and-paste geometry has proved to be a useful way of explaining how Old Babylonian scribes dealt with certain kinds of problems. Among these problems, the most important is one that has already been characterised, not without algebraic connotations, as a quadratic or second degree equation, because of the possibility of reducing it to an equation of this type.

In the textual corpus examined in this work, such kind of problem appears in IM52301, IM53965 and IM54559, where they come in two equivalent forms, represented in Fig. 2.1a, b. In the first one, a rectangle of known width is added to a

[3] The field is divided about which Akkadian verb in question here is: instead of *akālum*, this kind of multiplication may be given by *kullum*, to hold. See Sect. 3.2 for a technical discussion of the linguistic issue.

[4] There is not a complete consensus about the terms for square and cube roots. See Sect. 3.2 for the details.

[5] I adopt here a suggestion by Høyrup (2002, 31) that the hypothetical **bûm* given by CAD (B, 297) should in fact be *bāmtum*, obtained through a sequence of phonetic assimilations.

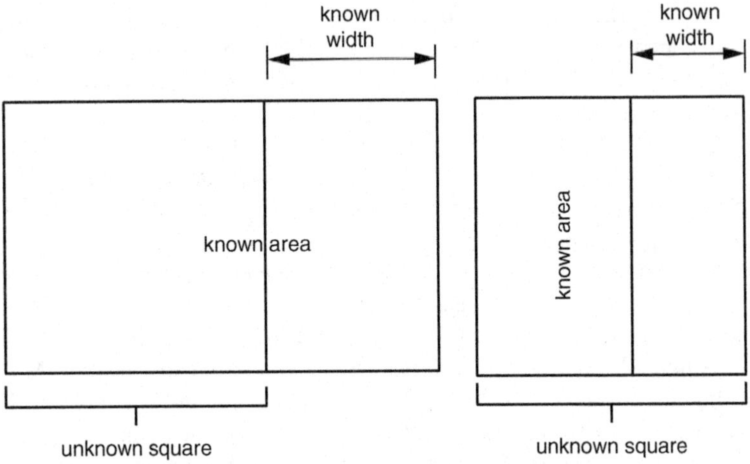

Fig. 2.1 (**a**, **b**) So-called problems about squares

square of unknown area (and side), resulting in a rectangle of known area. In the second form, a rectangle of known width is removed from a square of unknown area (and side), resulting in a rectangle of known area. In both cases, it is required to obtain the side of the square.

First of all, it should be noticed that the second case is equivalent to the first one. In order to see that, one can take the rectangle of known area in Fig. 2.1b and remove from it a rectangle with the same given known width, as in Fig. 2.2a, b. After this operation, one is left with a new unknown square to which a rectangle of known width is added, resulting in a known area, which is the situation in Fig. 2.1a.

Now let us get back to Fig. 2.1a and examine the strategy for solving the problem. As a first step, the known width is halved, and half of the corresponding rectangle is thought of as having been cut and pasted to another, convenient position. All is shown in Fig. 2.3.

The convenience of this step is revealed when one pays attention to the grey little square which has sides equal to half the known width, depicted in Fig. 2.4. As we can see, the hatched region equals the original known area (of a rectangle), so the new, bigger square involving it equals the original known area augmented by the grey square.

As a result, one is now enabled to compute the value of the side of the bigger square. From this, the side of the unknown square can be finally obtained by removing half of the known width, as in Fig. 2.4.

In the discussion of the individual tablets in Chap. 4, this procedure will appear again. Problems 1 and 2 of IM52301 present the situation as in Fig. 2.1b. Tablet IM53965 brings a problem following Fig. 2.1b too. Finally, in IM54559, both situations are perhaps simultaneously dealt with, that is to say, the one represented by Fig. 2.1a and the one represented by Fig. 2.1b.

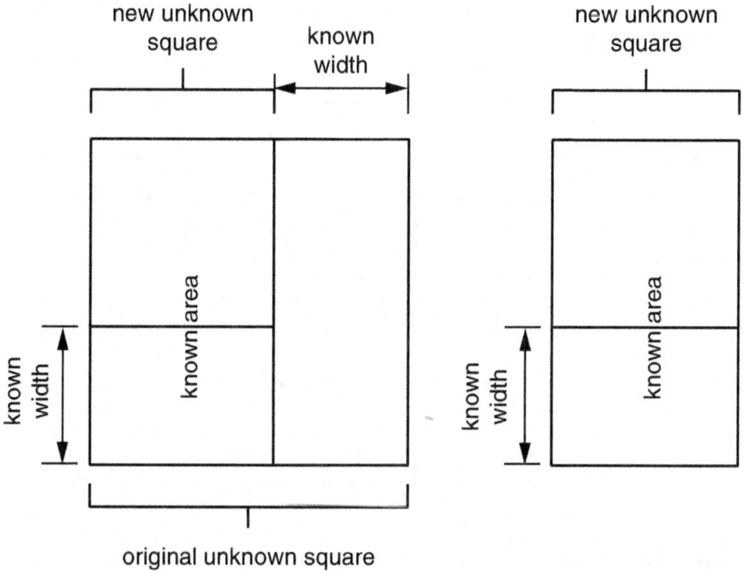

Fig. 2.2 (**a, b**) Reduction of one case of problems about squares to the other

Fig. 2.3 After cutting and pasting

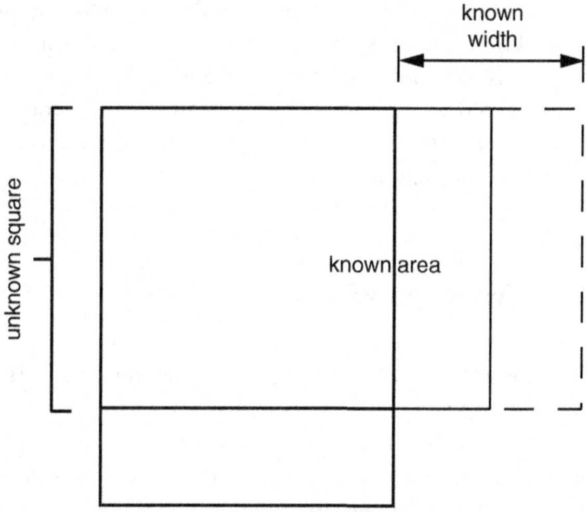

There are two other instances in the Tell Harmal tablets treated here that might be interpreted as related to cut-and-paste geometry, although they do not deal with problems about squares:

- In IM52301, as already mentioned, Problems 1 and 2 are efficiently interpreted if we resort to the previous procedures. Besides this, in the mathematical commentary to this tablet, I propose some preparatory steps for its interpretation that

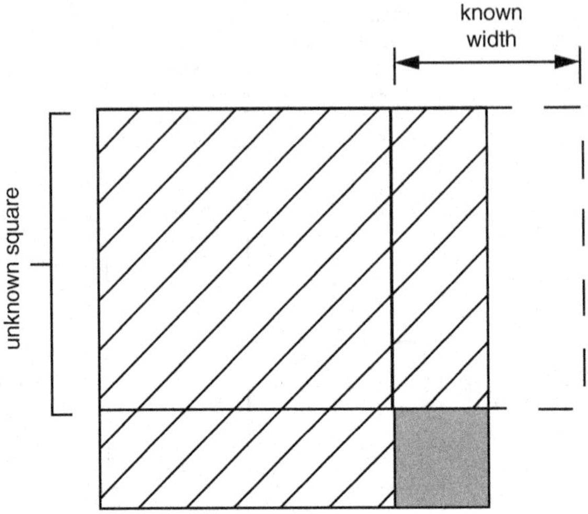

Fig. 2.4 The rationale behind cutting and pasting

rest upon making four copies of the quadrilateral that is given in the statement of the problem. In a certain way, making these copies reduces to copying and pasting, which might be seen as a variation of cutting and pasting.

- IM53957 is a tablet that contains a problem about an unknown quantity of barley. It is not a geometric problem in the traditional sense. Certain difficulties in the reading of some signs have led to two different interpretations. The second of these interpretations depends on a transformation of the data which is not explicitly given in the tablet. In my mathematical commentary, I propose a pair of schematic drawings that, although not exactly cut-and-paste geometry, has something to do with it.

2.3 Scaling of Figures

The scaling of figures is a frequent technique in Old Babylonian mathematical tablets. Høyrup (2002, 99–100) includes it in the more general concept of change of variable, although not to be understood as we do it in our study of functions and in our symbolic algebra. Old Babylonian "change of variable" is a more geometrical one, certainly difficult to represent, but having to do with transformations an unknown may suffer in order to make the solution of a problem easier to obtain. Scaling of figures is one of these transformations.

It should be noticed, however, that in no Old Babylonian tablet scribes make any explicit or theoretic reference to the technique of scaling. In spite of that, their activities are so consistently carried out in this respect, that we are compelled to assume their awareness of it. The next paragraphs discuss the presence of the scaling of figures in our 12 tablets.

The first analysed tablet, IM55357, is one of the best examples of a scaling of a figure. In it, the scribe systematically uses the fact that a right triangle can be scaled to an isosceles right triangle, maintaining the length of one of the legs. The scribe consistently keeps track of the changes of the values of the other leg and the area.

IM52301 contains two problems about quadrilaterals, a list of coefficients and a text written on the left edge that has been referred to as a third problem. In the mathematical analysis of Problems 1 and 2, the concept of scaling is extremely useful, especially in the preparatory steps. Apparently, the initial data of Problems 1 and 2 is transformed by means of a scaling, in order to rephrase them as problems about squares.

IM54478 deals with a cube of volume one and a half sar_v, of which the length of the edges is unknown. In order to determine this length, the scribe sets up another cube, a reference cube so to say. The scribe is then able to use the ratio of similarity between these two figures and determine what the problem requires. It is not clear whether the problem can be interpreted as a case of scaling of figures.

Unfortunately, IM53953 has some damaged parts that make it impossible to attain a complete understanding of its contents. The problem is about a figure with unknown dimensions, referred to as a triangle in the statement of the problem, but perhaps better understood as a certain trapezium formed inside a triangle. In order to obtain its dimensions, the scribe seems to set up another figure with proportional measures. This procedure can be understood as a scaling in an enlarged sense, because no measure of the figure to which the scaling is applied is known right from the start.

IM54538, IM53961 and IM53957 do not contain anything about the scaling of figures. IM54538, however, involves some knowledge of proportionality in the computation of the number of men necessary for a certain task.

IM54010 is a problem of unknown interpretation.

An interesting example of scaling is used in IM53965. In this tablet, a rectangle is scaled along its horizontal axis in order to produce a problem about squares of the type illustrated in Fig. 2.1a. In the end of the solution, the scaling is undone, which is an evidence that the scribe was able to control the effects of the scaling along the whole procedure.

As for IM54559, the interpretation of the problem is dubious, because of the bad state of its initial lines. However, if we are to believe in the proposed mathematical analysis, there is a step of the solution in which the scribe transforms a relation among three numbers into a pair of problems about squares, by raising the numbers to one third and then by raising the results to an unknown width. Had the scribe proceeded inversely, that is to say, had he raised the three numbers firstly to the unknown width, we would be assured that at this point he had three rectangles, on which a scaling of ratio one third would be subsequently applied. In the order the scribe applied the raisings, whether this counts or no as a scaling might be disputable. Anyway, it is interesting to contrast the possible "vertical" scaling here with the "horizontal" one in the previous problem. Both lead to problems about squares.

In IM54464, there are no figures, thus no scaled figures. However, it must be noticed how this tablet uses a clear sense of proportional numbers to solve the problem.

Finally, IM54011 is a straightforward problem about work assignments and a brick wall. There is no scaling of figures in it.

2.4 Systems of Measure

There are 12 units of measure, including the dubious readings, used in the tablets analysed in this work. Table 2.1 shows their Akkadian names, the corresponding logograms and their occurrences in the tablets. Each occurrence is indicated by LG, when it is written with a logogram; by SY, when the unit is written syllabically, or by N, when the unit is indicated by a special way of writing the numbers.

As the table shows, some units appear in the corpus only as logograms. This is the case of gur, še and gin_2, all units of capacity and weight. The cubit and the $sila_3$ are present in the corpus both with logograms and syllabically. Many of the remaining units never appear as a logogram: *mušarum*, *burum*, *nikkassum*, *qanûm* and *ašlum*. Next, the presence of ban_2 is indicated by a special way of writing the numbers. Finally, barig, if it is really present in IM53597 (see the corresponding commentary), should also be indicated by a special way of writing the numbers; however, in what seems to be a deviation from the norm, the numbers are written in the standard way in this case.

In the table, some of the occurrences are accompanied by a question mark. This is due to their dubious readings. Specifically, we cannot be completely sure that barig and gur really appear in IM53957 or that bur_3, cubit and *qanûm* appear in IM54010.

Finally, it should be added that Table 2.1 also brings three additional units which are not explicitly present in any of the examined tablets, but are of great help to explain their mathematical contents. They are the finger and the nindan, two length measures, and the sar_s, a surface measure.

In transcriptions, I always used the corresponding Akkadian terms, hopefully in the correct grammatical form (status and case).[6]

[6] Here it is necessary to make a more language-oriented commentary, which is important for the reader to know what she or he is going to read. In translations and commentaries, I would have liked to stick to some homogeneous pattern too. In my opinion, the names of the logograms would be the better option: $sila_3$, for instance. However, the very good translations given by grain, shekel, cubit, reed and rope were preferred over the Sumerograms and the Akkadian words. These translations are indicated in Table 2.1. The unit of volume sar is translated by the neologism sar_v (following the usage of other authors, as Robson (1999)). I considered it more adequate than keeping the original logogram sar, because the latter, in fact, does not appear in the corpus, in such a way that using it in translations might induce the reader to the mistake of thinking otherwise. Finally, in translations and commentaries, I maintained the Akkadian terms *parsiktum* and *nikkassum*.

Table 2.1 Units of measure present in the textual corpus

	IM55357	IM52301	IM54478	IM53953	IM54538	IM53961	IM53957	IM54010	IM53965	IM54559	IM54464	IM54011
Capacity: $sila_3 \times 10 \rightarrow ban_2 \times 6 \rightarrow barig \times 5 \rightarrow gur$												
qûm, $sila_3$							LG	SY			LG	
sūtum, ban_2											N	
parsiktum, barig							N?					
kurrum, gur							LG?					
Weight: (še = grain) × 3,0 → (gin_2 = shekel)												
uṭṭatum, še = grain											LG	
šiqlum, gin_2 = shekel											LG	
Volume: $nindan^2 \times cubit = sar_v$												
mušarum, $sar = sar_v$			SY									
Surface: $nindan^2 = sar_s \times 30,0$ → *burum*												
mušarum, $sar = sar_s$												
burum								SY?				
Length: (šu.si = finger) × 30 → ($kuš_3$ = cubit) × 3 → *nikkassum* × 2 → *qanûm* × 2 → nindan × 10 → (*ašlum* = rope)												
ubānum, šu.si = finger												
ammatum, $kuš_3$ = cubit						SY		LG?	SY			LG, SY
nikkassum												SY
qanûm = *reed*								SY?				
nindan												
ašlum = rope					SY							SY

As one can conclude from an examination of the table, units of measure are not extremely frequent in the studied tablets. As a matter of fact, there are four tablets (one third of the analysed corpus) in which no unit appears at all. In the cases where more than one unit appears in the same tablet, it makes sense to ask for their ratios. In IM53957, as already stated, there can be no assurance that barig, $sila_3$ and gur are really present, although it is possible to interpret the tablet by supposing that they are. This supposition takes into account the expected ratios 1 barig = 1,0 $sila_3$, 1 gur = 5,0 $sila_3$. On the other hand, in IM54464, 1 ban_2 and 5 $sila_3$ of oil are converted to 15 $sila_3$, thus confirming that 1 ban_2 equals 10 $sila_3$ (that is to say, $sila_3 \times 10 \rightarrow ban_2$); similarly, the problem makes use of the fact that 180 grains = 1 shekel (that is to say, 1 grain $\times$ 3,0 $\rightarrow$ shekel). As for IM54010, its bad state of preservation does not permit of a meaningful mathematical interpretation of its contents. Ratios between units of measure are represented in the table, together with the identification of the classes to which each unit belongs (Capacity, Weight, Volume, Surface, Length). As for the grammar of the units of measure, two remarks are worth mentioning:

- In IM54538, it seems that a grammatical correction might be necessary. A unit of measure, *mušarum*, seems to be used without the usual convention of the status absolutus.
- In both IM53961 and IM54011, two cubits are written *ši-ta am-ma-tim*, where the unit of length behaves like a counted item and thus appears in the status rectus. It is interesting to notice that the case in which the unit is used is always the accusative-genitive plural. It is therefore a construction similar to "number *ina ammātim*", in which the preposition *ina* introduces the length unit, but which is fully attested only outside the Old Babylonian dialect (see GAG, §139 I, and CAD A2, s.v. *ammatu*).

Finally, it must be said that the units of measure used in the 12 mathematical tablets examined in this work are in accordance with what is known about Old Babylonian units and systems of measure.[7]

2.5 Abstract and Measure Numbers

While examining mathematical texts, it is useful to distinguish between numbers that are accompanied by a unit of measure and those that are not. The former are usually called measure numbers; the latter, abstract numbers. The rationale of this distinction is the thesis that scribes, in general, do not perform arithmetical multiplications and inversions on numbers provided with a unit of measure. Instead, scribes would consult a metrological table and convert a measure to an abstract number, on

[7] The study of ancient metrology is a challenging task, because of the many regional and chronological variations. For a thorough exposition of the units and the systems of measure in Ancient Mesopotamia, see Powell (1987–1990). For the region of the Diyala, see Gonçalves (forthcoming).

which arithmetical operations could be performed. This position has been convincingly sustained and exemplified by Proust (2007), in the analysis of the school curriculum of Nippur.

The 12 tablets examined here may be regarded as a useful contribution to the issue.[8] Although they are on the whole consistent with the picture described in the previous paragraph, they also bring some examples of multiplications that are performed using measure numbers instead of abstract numbers. In this way, the tablets from Tell Harmal help us add some nuance to the separation between abstract and measure numbers.

In IM55357 and IM52301, there is no contrast between numbers with measures and abstract numbers, as there are no units of measure in the texts of these tablets.

As already pointed out by Proust (2007, 218ff), IM54478 is a nice example of the contrast between the numbers that appear accompanied by a unit of measure in a problem and the abstract numbers. In this problem, a cubic volume of one and a half sar_v is given. Immediately, it is transformed into an abstract number. This and the other numbers on the tablet are always used without units of measure, that is to say, they are always treated as abstract numbers. Interestingly, the final results are not presented with units of measure, what might be an indication either that abstract numbers could be used for final answers or that sometimes units of measure could be tacitly assumed.

IM53953 does not use units of measure, so all its arithmetical operations are made with abstract numbers.

IM54538 does not follow the convention that multiplications should be carried out with abstract numbers only. In a passage, the scribe raises "40 to a rope of length," that is to say, he raises an abstract number to a measure number.

Tablet IM53961 is a good example of the use of abstract numbers. In it, measures given in cubits are converted to abstract numbers, with which the calculations are then carried out. It is interesting to notice too that the final result is not converted back to cubit. Whether it is left as an abstract number or as a different unit of measure (namely nindan; see the corresponding mathematical commentary), it is difficult to know.

IM53957 deals with quantities of barley, expressed in $sila_3$, as well as perhaps gur and *parsiktum*. All operations are performed on abstract numbers.

IM54010 seems to present some units of measure, but its bad state of conservation inhibits a clearer analysis.

The most interesting technical aspect of IM53965 is the already mentioned scaling. In this tablet, arithmetics is carried out with abstract numbers, except in one passage where the scribe raises a cubit that was broken off to 30, that is to say, he raises a measure number to an abstract number.

IM54459 deals with no units of measure, so all operations are made with abstract numbers.

[8] See again Gonçalves (forthcoming) for the relation between abstract and measure numbers in the Diyala.

In IM54464, intense use is made of abstract numbers. For instance, in order to verify that the price of 1 $sila_3$ of lard equals two thirds of the price of 1 $sila_3$ of oil, the scribe simply raises 6 (the abstract number that corresponds to the price of lard) to 40, obtaining 4 (the abstract number corresponding to the price of oil). He knows that 40 here is meant to be two thirds, and he has complete control of the orders of magnitude.

In IM54011, it is possible to see the use of abstract numbers in action. One interpretation for the mathematical procedure assumes that the scribe converts measures given in cubits, *nikkassum*, and ropes to abstract numbers. After performing arithmetical operations on these numbers, the scribe obtains the desired results.

2.6 Floating Point Arithmetic and Orders of Magnitude

So far in this book, I have been writing numbers using the comma "," as digit separator and the semicolon ";" as the separator of the integral and fractional parts. This is in line with an understanding of the Mesopotamian numerical system used by many old and recent publications in the field. However, it must also be said that this procedure makes assumptions that have been no longer completely sustained by the research in the history of Mesopotamian mathematics.

The main point at stake here is whether one should or not assign defined orders of magnitude to abstract numbers in cuneiform texts, that is to say, whether it makes sense or not to insert the separator of integral and fractional parts.

A second important point is that, in Old Babylonian texts, there was no sign to indicate a vacant sexagesimal place. Sometimes this is explained by saying that there was no zero in the writing of numbers.

To understand what to do, we need to resume the discussion about abstract numbers, but now with the focus on the way they are written.[9] As already seen, abstract numbers convey a set of important characteristics of cuneiform mathematics, which are usually summarised by saying that these numbers not only are independent of units of measure but also that they are written in the so-called sexagesimal place value notation (SPVN).

However, one further characteristic of the writing of numbers should be examined here, because it has a special impact in the way cuneiform mathematical tablets are edited and read by historians of mathematics. This is the fact that in actual clay tablets, numbers in SPVN did not include any sign to separate their integral and fractional parts. So, for instance, and this may be specially stressful for the modern

[9] The theme has been the object of attention by the history of mathematics for quite a long time and there are, indeed, very good presentations of it, since the classic work by Thureau-Dangin (1932) to Friberg's article "Mathematik" in the *Reallexikon der Assyriologie* (Friberg 1987–1990) and the same author's Chapter 0 of his edition of the mathematical tablets of the Schøyen Collection (Friberg 2007a).

reader, the sign for 1 in a mathematical tablet forces us to consider as possible interpretations a series of numbers differing by a power of 60:

$$\ldots 1\times 60^{3} \quad 1\times 60^{2} \quad 1\times 60^{1} \quad 1\times 60^{0} \quad 1\times 60^{-1} \quad 1\times 60^{-2} \quad 1\times 60^{-3}\ldots$$

or using ";" as the integral and fractional separator and "," as the other separator for digits,

$$\ldots 1,0,0,0 \quad 1,0,0 \quad 1,0 \quad 1 \quad 0;1 \quad 0;0,1 \quad 0;0,0,1\ldots$$

A second example will show the impact of the absence of a zero on the possible readings of a number. For instance, if one has 7.30 written on a tablet, that is to say, a number composed of the sexagesimal digits 7 and 30, then some possible reading are:

$$7\times 60+30, \text{or} 7,30$$
$$7+30\times 60^{-1}, \text{or} 7;30$$
$$7\times 60^{2}+30, \text{or} 7,0,30$$
$$7\times 60^{-1}+30\times 60^{-3}, \text{or} 0;7,0,30$$
$$7\times 60^{6}+30\times 60, \text{or} 7,0,0,0,0,30,0$$

These are not only theoretical questions, for it has been noticed by the field that in some mathematical texts more than one interpretation is possible, without making the text lose any degree of correctness or coherence. In our corpus, this is the case of IM55357 (see the corresponding mathematical commentary for more details). In other words, in certain problems the orders of magnitudes are not fixedly defined, so that we can assign any order of magnitude to any number present in the problem. Frequently, after one of them is chosen all others are determined.[10]

Furthermore, it is thought these numbers operate under the laws of a floating point arithmetic, that is to say, operations are determined except by the order of magnitude (Proust 2013). Thus, for example, when we see a scribe raising (this is Akkadian for "multiplying" as explained above in the section about words for

[10] As a matter of fact, order of magnitude as an idea that depends upon the number of digits of a number is foreign to the SPVN. If necessary, a scribe would understand the order of magnitude of an number through the idea that it may correspond to a measure number. As this is done by means of the so-called metrological tables, our problem of order of magnitude turns into a Mesopotamian problem of assigning a position to a number in one of these tables. A much quoted tablet from the region of Ešnunna, Haddad 104 (al-Rawi and Roaf 1984), brings clear occurrences of this: for instance, in Problem 1, the initial data is given as "The procedure of a (cylindrical) silo. 5, one cubit, is its diameter." In this case, the abstract number 5 is given as the diameter, but this is not sufficient for one to know its order of magnitude, so the scribe writes that it corresponds to one cubit. In fact, in the metrological table of lengths, we have a line that runs as "1 cubit 5", and our problem of determining the order of magnitude of an abstract number (the number 5, as given in the problem) reduces, from the point of view of the scribe, to determining in the adequate metrological table which of the occurrences of this number is the one being dealt with.

arithmetical operations) 30 to 12 and obtaining 6, this may correspond to any of the following calculations (among infinite others):

0;30 raised to 12 gives 6
30 raised to 12 gives 6,0
30,0,0,0 raised to 12,0,0 gives 6,0,0,0,0,0,0
0;30 raised to 0;12 gives 0;6

Thus, when editing a mathematical text, the modern scholar has to choose between, on the one side, inserting the digit separators (as well as leading, intermediary and final zeroes), thus producing an anachronic version of Mesopotamian arithmetics but immediately legible by the reader unaccustomed with cuneiform mathematics, and, on the other side, letting the readers deal with the situation all by themselves, which may not be a good policy if communication is aimed for. This issue has led to a number of different opinions and approaches, as listed by Proust (2007, 74ff), through which specialists aimed at rendering their transliterations more faithful to the intentions of scribes. The question has been very recently balanced by Høyrup (2012), who reminds us that transliterations always involve a degree of interpretation; for instance, when we decide whether a sign is to be read as part of a syllable or as a logogram. Thus, the insertion of integral–fractional separators and the consequent determination of the orders of magnitude of numbers would not in themselves be practices that corrupt the purity of transliterations. They are never pure. Additionally, Høyrup suggests that in the majority of cases scribes did work with well determined orders of magnitude, so insisting in their total indetermination would miss the point.

All in all, the question is an important one and has helped the field of the history of cuneiform mathematics to gain insight into its object.[11]

In this book, numbers are written in two different ways, a Mesopotamian and a modern way, according to what emphasis is required. Thus, when no statement is needed about the order of magnitude of a number, I use dots to separate its digits, as 7.30 in the example above. On the other hand, when the order of magnitude makes difference and must be made explicit, I use the comma and semicolon system already described and insert the adequate zeroes.

As regards the presentation of the mathematical texts, the reader should bear in mind the following. I offer the transliterations and transcriptions without any indication of integral–fractional separation, while translations and commentaries, that are directed to a more general public, include the comma and semicolon system too.

For instance, 7.30 appears many times in the transliteration and transcription of Problem 2 of IM52301, without, as just mentioned, any formal indication of integral and fractional parts. In the corresponding translation and mathematical commentary, however, it also appears, depending on the requirements of the context, as 7,30, that is to say, as a 2-digit sexagesimal integer, or as 7;30, that is to say, as a sexagesimal number with one integer digit and one fractional.

[11] For more about this see also Proust (2013).

Another interesting example occurs in IM53965. In it the scribe makes the addition of 8.20 and 6.15. The result, 8.26.15, shows that he was aware of the different magnitudes of the involved numbers and that indeed 8.20 is to be thought of as 8,20,0.

2.7 Mathematical Coefficients

One of the types of mathematical tablets are the lists of coefficients, which contain constants to be used in specific calculations, such as the area of a figure or a work load. Eleanor Robson (1999) gives a detailed study about lists of coefficients and the way coefficients are used in mathematical tablets.

The last part of the reverse of IM52301 brings a small list of coefficients, containing six entries. One of these coefficients indicates the volume of wall one man is capable of building in 1 day (see Robson 1999, 93ff), and it is used in two other tablets of our corpus, IM53961 and IM54011.

Another coefficient, indicating the quantity of bricks of a certain size a man is capable of transporting over a distance of 1 nindan in 1 day, is used in IM54538. The problem, however, has not a clear complete interpretation.

Chapter 3
Conventions

What follows is a general exposition of the principles that guide Chap. 4, where the texts of the mathematical tablets are presented and commented. Section 3.1 deals with the general rules for transliteration, transcription and translation. Section 3.2 exposes some special, problematic cases pertaining to the mathematical language and the solutions that were adopted here.

3.1 General Rules

The presentation of each tablet is divided under the following headings:

- References, Physical Characteristics and Contents.
- Transliteration and Transcription.
- Philological Commentary.
- Translation.
- Mathematical Commentary.

The first heading brings information about the main publications regarding the corresponding tablet, especially about where it was first published and, when available, about published corrections and amendments.

After this, I present a transliteration of the text of the tablet. This transliteration is the result of my examination of the published transliteration (or transliterations), amendments and photographs of the tablet. When I was fortunate enough to have access to additional photographs, this is indicated. Here are the conventions I followed[1]:

[1] I followed Borger's practical guidelines for transliteration (MesZL, pages 242–243). For the rules governing the transliteration (as well as transcription and translation) of numbers, see Chap. 2, under the heading Floating Point Arithmetic and Orders of Magnitude.

C. Gonçalves, *Mathematical Tablets from Tell Harmal*, Sources and Studies in the History of Mathematics and Physical Sciences,
DOI 10.1007/978-3-319-22524-1_3

- Damaged signs are indicated with square brackets. Examples: [*na*] indicates the presence of sign *na*, but it is only partially damaged or can be identified from context; [x] indicates the room for one sign that cannot be identified; [xxx] indicates room for approximately three unidentified signs; [. . .] indicates room for an unknown number of signs; as I did not collate the tablets, my indications could not be extremely detailed, so I did not use the markers ⌜, ⌝, ° and so on to indicate damage on the corners of signs.
- Signs, damaged or not, of dubious identification are indicated with a superscript question mark.
- Parts of the transliteration that should be added to the tablet are indicated with < >.
- Parts of the text that the scribe should have omitted are indicated with << >>.
- Corrections on the scribe's text could be indicated using <<A>><B>, meaning that sign B should be substituted for sign A in the text on the tablet; however, this gets rather heavy to read. Thus, to indicate the above replacement, I simply use B!(A).
- When there is more than one possibility for establishing the transliteration, these possibilities are written inside curly braces, and they are separated by one or more slashes. Thus, in IM52301, line 17, {20/min$_2$} indicates two possible ways of reading the cuneiform sign.
- Logograms are always transliterated in small letters, regardless of whether their pronunciation is known or not.

The transcription that accompanies the transliteration is my proposal of how a scribe would read, in Akkadian, the text of the tablet. Besides, the transcription has the goal of offering the reader a meaningful and complete grammatical interpretation of the text, as regards my readings not only of the logograms but also of the grammatical forms and syntactical relations. The transcription adopts the following conventions:

- Parts corresponding to readable signs of which I could not make sense are indicated with [. . .].
- Multiple possibilities for transcribing a part of the transliterated text are written within curly braces, each possibility separated by a slash. Thus, for instance, in line 3 of IM55357, uš is transcribed as {*ēmidum*/*rēdûm*}.
- Hyphens separating the enclytic particle -*ma* are maintained in the transliteration, in order to emphasise their syntactical role. Possessive and oblique personal suffixes, on the other hand, appear directly after their nouns and verbs, without the hyphen.
- In the transcription, I inserted some stops, in order to make more evident my interpretation of the text. Commas, semicolons or colons are absent.

Finally, I have adopted the following two grammatical conventions that affect transliterations.

There are some cases where there is no consensus among specialists as to whether a vowel should be regarded as long by nature or long as a result of a vowel contraction, that is to say, whether it should carry a macron or a circumflex. In the analysed tablets, these are the instances that occur: *kī*, *šū* (GAG, AHw) or *kî*, *šû* (CAD). In transcriptions, in this work, *kī* and *šū* were preferred over *kî* and *šû*. This does not represent any particular stance on the grammatical issue, but only the necessity of sticking to one and only form throughout the text.

The second grammatical convention has to do with the interrogative pronoun *mīnum* or *minûm*.The usual Babylonian form is *minûm* or *minû* (GAG §47b; AHw 655). The normal Assyrian form *mīnum* changes to *minûm* in interrogative sentences as a result of the sentence modulation (GAG *ib.*). In the transcriptions, I always employed *mīnum*, as the lengthening of the final vowel is never explicitly attested by the cuneiform of these texts. In this respect, writings like *mi-nu-um* can be read in both ways, but *mi-nu*, as in IM52301, almost certainly points to *mīnu* and not *minû* (supposedly to be written with an additional final u-sign, that is to say, *mi-nu-*u_x; but even this must be taken with a grain of doubt, for scribes were not always consistent).

The third subhead in the presentation of a tablet is the philological commentary. Here, in general, I bring all the variant readings present in the published literature, except when a variant would not necessarily lead to a different reading or interpretation. For instance, if I write *ba*$^?$ in a transliteration, and some other author writes simply *ba* in their publication, this is not considered a divergence and consequently is not indicated in the philological commentary. Differences due to ambiguity in signs with phonetic values involving the vowels "e" and "i" were not considered meaningful either, so they are not indicated here: for instance, *e-pe*$_2$*-ši-ka* and *e-pi*$_2$*-ši-ka*. However, if this ambiguity might lead to the identification of different grammatical forms, the divergence is indicated: for instance *ḫe-pe*$_2$, *ḫi-pe*$_2$, *ḫi-pi*$_2$ and *ḫe-pi*$_2$, variants that according to one's point of view might indicate either the imperative or the stative of *ḫepûm*. Finally, the philological commentary also brings grammatical remarks that I considered necessary or helpful in the understanding of the original text.

The translation tries to respect the lexical meaning of the Akkadian words, when this meaning is known. I also tried to maintain the same word order of the Akkadian text. When this would lead to too much stress on the reader, I changed the word order. The frontier between the two cases is a blurred one. For instance, in my opinion, the position of the Akkadian verb at the end of the sentence can be maintained in segments of text like "two thirds and two thirds cause them to combine, and 26 40 you see" (IM53957, lines 7 and R1), but would sound a little too odd in "A reed I took and its size I do not know", instead of "I took a reed and I do not know its size" (IM53965, lines 2 and 3). All in all, my translations should not be considered conformal ones, in the senses used by Høyrup (2002, 40ff; 2010) and Friberg (2007a, 2), but the idea of a conformal translation is of course an inspiration and a deep influence. Parts of the transcription that are multiple readings are multiply translated. Thus, {*ēmidum/rēdûm*} is translated {leaning/following}.

Finally, comes the mathematical commentary, where a possible technical explanation is advanced for each problem. I systematically avoided using symbolic algebra, in order to be as close as possible to what the field presently thinks would be the cognitive devices available to the scribes. Instead, the reader will find the mathematics of the tablets explained in plain words and cut-and-paste diagrams. In order to facilitate the understanding of the mathematical commentary, some expressions that are alien to Old Babylonian mathematics are used: numbers, fractions, ratios, geometrical figure, problem, solution. They should be regarded by the reader as helpful metalanguage and not as the very contents of the analysed tablets. On the other hand, a number of words and expressions specific of Babylonian mathematics are present in the translations and the mathematical commentaries: to raise a number to another, to accumulate and to remove, as explained in Chap. 2, in the section about words for Arithmetical Operations. Their presence should by no means be considered an obstacle to the understanding of the contents of the mathematical tablets, but rather as a reminder that we are dealing with mathematical concerns and practices that have traits of their own.

3.2 Special Mathematical Cases

The following paragraphs tackle specific problems encountered in the making of this work and the solutions that were adopted. These problems have to do with the way scribes used Akkadian, as well as some Sumerian background, in order to compose mathematical texts. Thus, from a specific point of view, these are language problems. However, as these problems owe their existence to the need of expressing mathematical ideas, they are not independent from mathematics and, in this way, they belong together.

As the reader will see, besides quoting the specialised literature on the history of Mesopotamian mathematics, I also quote the available comprehensive dictionaries of Akkadian, that is to say, Wolfram von Soden's *Akkadisches Handwörterbuch* (AHw) and *The Assyrian Dictionary of the Oriental Institute of the University of Chicago*, also known as *The Chicago Assyrian Dictionary* (CAD). AHw was published from 1965 (Volume I) to 1981 (Volume III), and CAD's volumes began to appear in 1956 (Letter Ḫ). While CAD's last volume (Letter U/W) came to light only in 2010, the greater part of its 21 volumes was conceived and published before the present interpretation on Mesopotamian mathematics and its vocabulary took shape (see Høyrup (1996) for a history of the research on the field). As a consequence, both AHw and CAD reflect a period where the nuances of terminology that I present here were not perceived, a reason by which these dictionaries should not be taken as authorities, but as additional references for those that want to form an opinion on the mathematical vocabulary of cuneiform tablets.

3.2.1 *The Term for Reciprocal: i-gi and* **igi**

This word is usually written with the Sumerian logogram igi, corresponding to the Akkadian *igûm*. In the textual corpus I analysed here, however, igi occasionally receives a syllabic writing, *i-gi*.

This term is commonly translated as "reciprocal". However, the way its meaning is formed is not known. This is due to two reasons: firstly, there is no general agreement about the etymology of the word; secondly, the term seems to have been used only exceptionally outside mathematical contexts, so that it is not possible to contrast its mathematical meanings and usages with its non-mathematical ones, if they indeed existed (see Høyrup 2002, 27–30, for a detailed discussion).

Much as "reciprocal" is an acceptable and convenient translation, it carries along ideas from the semantic field of "reciprocal", as present in our arithmetic, which are very extraneous to Old Babylonian arithmetic. For instance, the idea of the inverse operation of the multiplication and the existence of a reciprocal for every given non-zero number. In order to keep this difference in mind, this term is kept untranslated in this work.

3.2.2 *The Sumerian Logogram for "Triangle":* **sag.du$_3$** *or* **sag.kak?**

The pair of cuneiform signs that make the composite logogram for triangle poses a problem for transliteration, once its Sumerian pronunciation is not known with certainty. It might have been read in any of the ways listed above. Among specialists, one can discern the following usages:

- Friberg (2007a, b): sag.kak
- Robson (1999), Høyrup (2002), Proust (2007): sag.du$_3$

The present work uses sag.du$_3$.

3.2.3 *The Akkadian Word for Triangle: sattakkum and santakkum*

Both forms are attested. Yet, transliterations and transcriptions should, whenever possible, point to an orthographic homogeneity of the corpus. Because of this, *sattakkum* is better than *santakkum* in the case of Tell Harmal, once the former spelling is the only one employed explicitly by our scribes, namely in tablets IM55357 and IM53953 (each tablet from a different Ešnunnan linguistic variety, as explained in Sect. 5.5). I also take it to be the Akkadian equivalent to sag.du$_3$.

3.2.4 *How Should* **sag.du$_3$** *and sattakum Be Translated: Triangle, Wedge or Peg-Head?*

Friberg (2007a, b) translates sag.kak as "peg-head (triangle)". Robson (2007, 100) translates this composite logogram as "wedge". Høyrup (2002) and Proust (2007) use "triangle", also accepted here.

3.2.5 *A Term for Multiplication: šutākulum or šutakūlum?*

Thureau-Dangin (TMB, 219) assumes it to be *šutakūlum*, that is to say, the Št-Stem of the verb *kullum*, to hold, but Neugebauer and Sachs (MCT, 159) take it to be *šutākulum*, that is to say, the Št-Stem of the verb *akālum*, to eat.

In the more recent literature, positions vary too:

- Kazuo Muroi sustains that the verb from which the form is correctly derived is *akālum*: "In mathematical texts, the word *šutākulum*, 'to make (them) eat one another', which seldom occurs in non-mathematical contexts, is used in the meaning of 'to square, to multiply'" (Muroi 2003, 254).
- Eleanor Robson, in a 2001 paper, while explaining the derivation of the term *takiltum*, comments that "the OB verb 'to multiply geometrically' … is probably the causative reciprocal form of *kullum* …" (Robson 2001, 191). On the other hand, in a later publication, the form from the verb *akālum* is employed: "… *šutākulum* 'to combine', indicates a square constructed in the course of a cut-and-paste procedure" (Robson 2008, 113).
- Jens Høyrup interprets it as "*šutakūlum*, 'to make [two segments *a* and *b*] hold each other', *viz*., as the sides of a rectangle …", therefore as a form of the verb *kullum* (Høyrup 2002, 23).

The very fact that the field has not reached a consensus should be an indication that more evidence is expected before closing the question.

I assume the term comes from the verb *akālum*, and my main reason for it is a writing *tu-uš-ta-ak-ka-al-ma*, that appears twice in tablet YBC4675, in lines 12 and R15 (MCT, 45). The doubling of the *k* can only be possible with the verb *akālum*, and *tu-uš-ta-ak-ka-al-ma* is normalised as *tuštakkalma*, a form of the present of the Št-Stem (see also Kouwenberg 2010, 408). Once, these are isolated writings, I agree that there is still room for further discussion.

3.2.6 *Placing Aside a Number for Later Use:* (x) *rēška likīl*

This expression was translated by Thureau-Dangin (TMB, 224) as "que ta tête retienne (tel nombre)", that is to say, "may your head hold *x*". In this interpretation, head (*rēšum*) is the subject of the sentence and it is it that holds (*likīl* is a precative form of *kullum*, to hold) the number. Høyrup supports the same position, noticing additionally that the expression "seems to be reserved for numbers that are not to be inserted in a fixed scheme", in contrast to another way of placing numbers in calculations (2002, 40).

However, according to Muroi (2003) and Friberg (2007a, 337; 2007b, 90), this expression corresponds instead to "let it hold your head", that is to say, the number x and not *rēška* is the subject of *kullum*. Thus, the number holds the head, as it may hold today a position in the memory of our pocket calculators.

This second interpretation is accepted by the present work. However, it must be clear that grammar does not enable us to distinguish whether it is the number or the head that holds the other. *rēška* is the status constructus of *rēšum* (head) with the suffix for the second singular person, "your head". The nominative and the accusative are indistinguishable in this form.

In the analysed tablets, this phrase appears in the following places:

- Once in Problem 1 of IM52301, to reserve the number 1,46,40. The scribe continues with other calculations and afterwards uses this number again.
- Twice in Problem 2 of IM52301, to reserve first the number 7,30 and then the number 7;30. In order to recover 7,30, the scribe writes "7,30 that holds your head".[2] However, 7;30 is recovered with a different expression, *takiltum*, to be examined shortly.
- Once in IM53965, where the reserved number is 8,20, that, after some calculations, is simply put back to use, without any special expression, as in Problem 1 of IM52301.

3.2.7 A Term for Recovering a Reserved Number: takiltum (AHw), tākiltum or takīltum (CAD)?

There is no consensus among specialists in relation to the correct Akkadian transcription of this term. On the one side, according to Robson, the correct writing is *tākiltum*, and the term is "a noun derived from the verb of geometrical squaring *šutākulum*" (Robson 2008, 113). On the other side, Jens Høyrup, following Neugebauer (MCT) and von Soden (GAG, paragraph 56), sustains the correct spelling is *takīltum* (Høyrup 2002, 23; 2010, 401), and the "term can only derive from *kullum*" (Høyrup 2002, 23). So does Kazuo Muroi saying that "it is derived from the verb *kullum*" (Muroi 2003, 254), although his reasons are different from Høyrup's. In the same line, Friberg (2007b) transcribes the word, as present in Plimpton 322, with the spelling *takīltum* (Friberg 2007b, 90). Finally, both the AHw and the CAD link the word to the verb *kullum*.

The issue is maybe not independent from the *šutakūlum versus šutākulum* question just described. I do not exclude the possibility that *šutakūlum* (written in this way, thus deriving from *kullum*) might be a link between *kullum* and *takīltum*. However, I refrain from taking this for sure, mainly because of the above mentioned

---

[2] Here too grammar does not enable us to distinguish this sentence from "that your head holds".

spelling *tu-uš-ta-ak-ka-al-ma* that seems to point rather to *šutākulum* (from *akālum*) as a correct form.

In the present work, I assume the word comes from the verb *kullum*. Consequently, I use the spelling *takīltum*. However, I do not proclaim here any position as to the way the word is derived from the verb.

In relation to its meaning and translation, positions also vary. It has been translated as "the *tākiltum*-square" (Robson, *ib*) and "the made-hold" (Høyrup, *ib*). Muroi sustains that its meaning is literally "the one which contains (something)" (Muroi 2003, 262) and Friberg translates it as "the holder" (Friberg, *ib*). The AHw writes "bereitstehende (Verfügungs-)Zahl", whereas the CAD reports only that it is a mathematical term. In order to emphasise the strangeness of the term in relation to our mathematics, I leave it untranslated in the present work: *takiltum*.

In our corpus, it is used only once, in the solution of Problem 2 of IM52301. After raising 10 to 0;45, the scribe obtains 7;30. He then asks us not to forget this number or to put it aside for later use. The expression he employs for this is analysed above, *rēška likīl*. He then goes on making his calculations, until a point where he needs to use 7;30 again, where he writes: "Cut off 7;30, your *takiltum*, from 22;30". Thus, 7;30 was put aside with *rēška likīl* and then recovered with *takiltum*.

3.2.8 Terms for Square and Cube Roots

In the examined corpus, the vocabulary to deal with square and cube roots comprehend the Sumerian $ib_2.si_8$ and ib_2.si.e, and the Akkadian *basûm* and *maḫārum*. In other mathematical texts, it is also common to find ba.si, which is the Sumerian term that originates the Akkadian *basûm*.

The verb *maḫārum* means to accept, and its Št-Stem, to confront, is used in mathematical texts to indicate the construction of a square. The derived noun, *mitḫartum*, may refer either to the square or its side, and it is best translated as "confrontation" (Høyrup 2002, 25). In order to exemplify how it appears in mathematical texts, I will take IM54472. This tablet is also from Tell Harmal, but it is not dealt with in the present work, for having been published without copy or photograph, by Bruins (1954), it is not possible to make even a minimal checking of the reading.[3]

In IM54472, the scribe gives the area 26,0,15,0 of a square region and asks for the *mitḫartum*[4]:

26.15 a.$ša_3$ […] *mitḫartī kīa* […] (lines 2 and 3)

---

[3] The tablet is indeed badly published. The numbering of the lines is inconsistent and, without being able to check the reading, it is not possible to be sure of the unusual a.ša and ib.si.e that Bruins writes instead of the more common a.$ša_3$ and ib_2.si.e. Here I use the latter forms, but collation is necessary.

[4] Here I simplify the issue of the orders of magnitude, for my purpose is to deal with the vocabulary of square and cube roots. In the text of the problem, we have the abstract number 26.15. The writing $26, 0, 15, 0 = 26 \times 60^3 + 0 \times 60^2 + 15 \times 60 + 0$ is only a possible interpretation for it. Although there was no notation for 0 in Old Babylonian mathematics, scribes were able to tackle empty sexagesimal positions.

26,0,15,0 is the area. How/What is my confrontation?

After a sequence of calculations, the answer is obtained:

39.30 *mitḫartaka* (line10)
Your confrontation is 39,30

In fact, 39,30 times 39,30 gives 26,0,15,0.

In the middle of the text, where the scribe is carrying out auxiliary calculations, he needs the square root of 15,0, which is 30 in the sexagesimal base. He then writes:

15 *mīnam* ib_2.si.e 30 ib_2.si.e (line 5)

As we see, the scribe is expressing himself differently here. I will postpone the translation a few paragraphs. For now, let us pay attention to the following points. It is possible to suggest that *mīnam* is the object of a verbal construction. In this case, ib_2.si.e would have verbal meaning, and the sentences could be translated: "What (object) does 15 (subject) ib_2.si.e (as a verb)? It (referring to 15) ib_2.si.e (as a verb) 30". In the same problem, the pattern appears once more, when the scribe needs the square root of 1,44,1, that is 1,19:

1.44.1 *mīnam* ib_2.si.e 1.19 ib_2.si.e (lines 7 and 8)

However, ib_2.si.e can also be used as a noun. Again IM54472 brings an example of this usage:

1.19 *ana* 30 ib_2.si.e *išī-ma* (lines 8 and 9)
Raise 1,19 to 30, the ib_2.si.e

Thus far, we have seen that IM54472 uses *mitḫartum* to refer to the side of a thing, namely the given square area, and uses ib_2.si.e when making calculations. A very similar pattern is found in one of the tablets analysed in the present work, IM54478. In it, the scribe deals with a hole excavated in the earth. The hole has the shape of a cube, a fact stated by the scribe with the following words:

mala uštamḫiru ušappil-ma (line 2)
As much as I caused it to confront itself, so I excavated

Thus, a form of the verb *maḫārum* is used to describe the superficial, or horizontal, shape of the hole, which is a square, while the verb *šapālum*, to excavate, refers to its vertical component. In the auxiliary calculations, the scribe resorts to a form similar to ib_2.si.e. Here it is:

7.30 *mīnam* ib_2.si_8 30 ib_2.si_8 (lines R1 and R2)
What (object) does 0;7,30 ib_2.si_8 (as a verb)? It (referring to 0;7,30) ib_2.si_8 0;30

In this case, instead of the computation of a square root, we have a cube root. In fact 0;30 times 0;30 times 0;30 gives 0;7,30. What is interesting here above all is that although 0;30 is the numerical cube root, it refers directly only to the confrontation and not to the depth of the hole, as we can see in the final answer:

30 *mitḫartaka* 6 *šupulka* (lines R5 and R6)

0;30 is your confrontation. 6 is your depth

Thus far, the examples show a tendency for *mitḫartum* to refer to the side of a square thing, as a square region (IM54472) or the upper surface of a hole in the earth (IM54478). Yet, the square thing may also be an abstract one, as in BM13901, a much quoted tablet dealing with problems about squares (MKT Vol. III, 1ff; Høyrup 2002, 50ff).

On the other side, ib_2.si.e or ib_2.si_8 is used in auxiliary calculations to obtain numerical square and cube roots. The reader will notice that in the tablets analysed in this book, ib_2.si.e appears exclusively in the context of the square root (IM53953, IM53965 and IM54559, to which we may add the just mentioned IM54472). However, a correlate to ib_2.si.e, which is ib_2.si, may also be found in association with the cube root, as attested by IM90889, a table of cube roots from Tell es-Seeb, also in the region of the Diyala (Isma'el and Robson 2010, 157–158). As for ib_2.si_8, it is used to obtain the cube root in IM54478, as we have just seen, and also the square root, as we will see in IM55357.

All in all, when doing the calculations, there seems to be no strict specialisation of square and cube root terms. This may be associated with the way Mesopotamian scribes understood areas and volumes:

- 1 sar of area is equivalent to a square of sides 1 ninda
- 1 sar of volume is equivalent to a right square prism with base 1 sar of area and height 1 cubit (we need 12 cubits to make 1 ninda)

From this, two characteristics are salient. Firstly, areas and volumes are measured with the same units (for instance, the sar). In the second place, it comes the very important feature that the unit of volume is not a cube: an area is converted to a volume by the addition of the standard thickness of 1 cubit. Thus, the cube in particular has not in Mesopotamian mathematics the same epistemological centrality that it has in Greek mathematics. We are dealing here with a very different spatial experience. We could also say it by noticing that while in our metrology the area and volume units arise from the appropriate use of a same length (the side of the square and the edge of the cube), in the Mesopotamian system they are built from a same square (a "confrontation"): a volume is a square with thickness.[5]

Thus, both square and cubic things have a confrontation. In the square, the confrontation is the side; in a cube, it is the side of the upper (or the bottom) face. It is only with the employment of the sexagesimal system that this side can be made numerically equal to the height of the cube. IM54478 illustrates this: 0;30 is the confrontation, the result of the numerical operation performed by ib_2.si_8, while 6 is the depth, as we have just seen.

Having said all this, it is now necessary to deal with two remaining questions:

- How should we translate ib_2.si.e and ib_2.si_8?
- Do they have Akkadian equivalents?

[5] Which is similar to the thought that "transforms a 'Euclidean line' in a 'broad line'" (Høyrup 2002, 51), by interpreting a given line *l* as a rectangle of sides *l* and 1.

One complication arises from the fact that si_8 and sa_2 are two values of the same cuneiform sign, so that ib_2.si_8 might be ib_2.sa_2 instead. From this, Attinger (2008) raises the hypotheses that we would be dealing in fact with two different terms: ib_2.si and ib_2.sa_2. However, the evidence from the Diyala shows that both forms are used in the very same contexts, from what I assume that, at least in this region, we are dealing with ib_2.si_8 and a variant writing ib_2.si.e. As for translation, I agree that the term dwells in the semantic field of "being equal", as other authors do too (Høyrup 2002; Friberg 2007a). In this work, ib_2.si.e as a verb is translated as "to make something equal". As a noun, it is translated as "the equal". Thus, the passages from IM54472 become:

1.44.1 *mīnam* ib_2.si.e 1.19 ib_2.si.e
What does 1,44,1 make equal? It makes 1,19 equal
1.19 *ana* 30 ib_2.si.e *išī-ma*
Raise 1,19 to 30, the equal

As for Akkadian equivalents, there is nothing in our corpus that points to their existence. The choices scribes made between the terms derived from *maḫārum* or ib_2.si.e and ib_2.si_8 point perhaps to a usage of the former to refer to square things, while the latter is restricted to arithmetical contexts. There is no indication that they could be used interchangeably or that they simply translate each other.

Finally, the reader should be reminded that these indications are rather tendencies than rigid rules that scribes followed, for which the available evidence is so fragmentary and sparse to constitute a proof. Furthermore, one should take into account that there is a possible regional variation of usage (see again Attinger (2008) for a discussion on that).

3.2.9 *A Text Structuring Expression:* za.e ki_3.ta.zu.ne/*atta ina epēšika/You, in Your Doing*

The Sumerian expression za.e ki_3.ta.zu.ne, corresponding to the Akkadian phrase *atta ina epēšika*, is usually translated as "you, in your doing". It was used in mathematical problems to mark the end of the statement and the beginning of the solution. The scribes seemed to mean that "in order to solve the problem, you have to do" this and that. The expression carries, in this way, a connotation of purpose and intention.

As for the Sumerian writing, the expression seems to be composed by the elements:

- za.e, the personal pronoun you
- ki_3.ta.zu.ne which is a writing of the verbal form made up of the grammatical elements ak+ed+zu+ne, pronounced approximately as kedazun(n)e[6]

[6]This is not the place to enter into details about Sumerian pronunciation, but the reader should be reminded that a same cuneiform sign may be employed with different phonetic values. In particu-

In the Tell Harmal tablets that I analyse here, the expression occurs in Sumerian in the first two texts, IM55357 and IM52301, while the remaining ten contain the Akkadian version. This might be an additional evidence of the split between, on the one side, IM55357 and IM52301 and, on the other side, the group of those ten tablets. The split may point to an earlier composition of the first two tablets in relation to the others, for which a language with less Sumerian elements is one of the defining characteristics.

However, while the Akkadian expression is written exactly in the expected way, that is to say, in the same way that it occurs in the remaining Mesopotamian mathematical corpus, the Sumerian expression in IM55357 and IM52301 deviates from the formula stated above. This will be dealt with in the philological commentaries of these tablets. In both cases, there are indications that the scribes tried to reproduce the expected sound but with a different arrangement of cuneiform signs.

lar, the sign ak, that corresponds to the Sumerian verb "to do", may be read ak, aga, a_5, but also, ki_3, ke_3 and kid_3. For more about the Sumerian verb ak, "to do", see Powell (1982) and Attinger (2005).

Chapter 4
Mathematical Tablets

4.1 IM55357

4.1.1 References, Physical Characteristics and Contents

Originally published by Baqir (1950a), this tablet drew the attention of scholars and arose a debate about its significance. The central point of this discussion was whether this mathematical problem and its solution could be taken as a proof that Old Babylonian scribes knew the laws of the similarity of triangles. Drenckhahn published his analysis of this text both in *Sumer* (1951) and, with a new transliteration, in the *Zeitschrift für Assyriologie* (1952a). Bruins (1951) published a very short comment on it, together with a letter to the editor of *Sumer*, objecting to some of Drenckhahn's conclusions, and a note reinforcing these objections. This created occasion for Drenckhahn to reply Bruins's statements in a new letter to the editor (1952b). A few years later, the tablet was again commented on by Bruins (1955). Almost half a century went by before Høyrup (2002) produced a new full transliterated text of the tablet, together with a translation and a mathematical interpretation. For the following transliteration, I also had access to a pair of colour photos of the obverse of the tablet, lacking, however, a portion of the right-side part and the right and the bottom edges, where part of the text is written.[1]

The tablet was found in room 301 (Hussein 2009, 92) of Tell Harmal, at the stratigraphic Level III (Baqir 1950a, 39), which corresponds to the time of Ipiq-Adad II and the beginning of the First Dynasty of Babylon. It measures 9.5 × 6 × 3 cm.

[1] I am indebted to Hermann Hunger for allowing me to have access to these photos.

C. Gonçalves, *Mathematical Tablets from Tell Harmal*, Sources and Studies in the History of Mathematics and Physical Sciences,
DOI 10.1007/978-3-319-22524-1_4

4.1.2 Transliteration and Transcription

Obverse

(1) sag.du$_3$ 1 uš 1.15 uš gid$_2$ 45 sag.ki an.ta
sattakkum. 1 *šiddum.* 1.15 *šiddum arkum.* 45 *pūtum elītum.*
(2) 22.30 a.ša$_3$ til *i-na* 22.30 a.ša$_3$ til 8.6 a.ša$_3$ an.ta
22.30 *eqlum gamrum. ina* 22.30 *eqlim gamrim* 8.6 *eqlum elûm*
(3) 5.11.2.24 a.ša$_3$ uš 3.19.3.56.9.36 a.ša$_3$ 3.kam
5.11.2.24 *eqlum* {*ēmidum/rēdûm*} 3.19.3.56.9.36 *eqlum šalšum*
(4) 5.53.53.39.50.24 a.ša$_3$ ki.ta
5.53.53.39.50.24 *eqlum šaplûm.*
(5) uš an.ta uš murgu$_2$ uš ki.ta *u*$_3$ *mu-tar-ri-it-tum mi-nu-um*
šiddum elûm {*šiddi būdim/šiddum warkûm*} *šiddum šaplûm u muttarrittum mīnum.*
(6) za.e ki$_3$.ta.zu.un.ne igi 1 uš du$_8$.a *a-na* 45 il$_2$
atta ina epēšika igi 1 *šiddim puṭur. ana* 45 *iši.*
(7) 45 igi.du$_3$ 45 nam 2 il$_2$ 1.30 igi.du$_3$ 1.30 nam 8.6 a.ša$_3$ an.ta
45 *tammar.* 45 *ana* 2 *iši.*1.30 *tammar.* 1.30 *ana* 8.6 *eqlim elîm*
(8) il$_2$ 12.(9) igi.du$_3$ 12.9 a.na(ba?).am$_3$ ib$_2$.si$_8$ 27 ib$_2$.si$_8$
iši. 12.9 *tammar.* 12.9 *mīnam* ib$_2$.si$_8$. 27 ib$_2$.si$_8$
(9) 27 sag (erasure) 27 *ḫi-pe*$_2$ 13.30 igi.du$_3$ igi 13.30 du$_8$.a
27 *pūtum.* 27 *ḫipe.* 13.30 *tammar. igi* 13.30 *puṭur.*
(10) nam 8.6 [a.š]a$_3$ an.ta il$_2$ 36 igi.du$_3$ uš gaba uš 45 sag.ki
ana 8.6 *eqlim elîm iši.* 36 *tammar šiddam miḫrit šiddim* 45 *pūtim.*
(11) *na-as*$_2$*-ḫi-ir* uš 27 sag.du$_3$ an.ta *i-na* 1.15 ba.zi
nasḫir. šiddam 27 *sattakkim elîm ina* 1.15 *usuḫ.*
(12) 48 ib$_2$.tag$_4$.a igi 48 du$_8$.a 1.15 igi.du$_3$ 1.15 nam 36 il$_2$
48 {*izzib/iriaḫ*}. *igi* 48 *puṭur.* 1.15 *tammar.* 1.15 *ana* 36 *iši.*
(13) 45 igi.du$_3$ 45 nam 2 il$_2$ 1.30 igi.du$_3$ 1.30 nam 5.11.2.24 il$_2$
45 *tammar.* 45 *ana* 2 *iši.* 1.30 *tammar.* 1.30 *ana* 5.11.2.24 *iši.*
(14) 7.46.33.36 igi.du$_3$ 7.46.33.36 a.na(ba?).am$_3$ ib$_2$.si$_8$
7.46.33.36 *tammar.* 7.46.33.36 *mīnam* ib$_2$.si$_8$.

Edge

(E15) 21.36 ib$_2$.si$_8$ 21.36 sag.ki <sag.>du$_3$ 2.kam$_2$
21.36 ib$_2$.si$_8$. 21.36 *pūt sattakkim šanîm.*
(E16) *ba* 21.36 <*ḫi*>*-pe*$_2$ 10.48 igi.du$_3$ igi 10.48 du$_8$.a.
ba 21.36 *ḫipe.* 10.48 *tammar. igi* 10.48 *puṭur.*
(E17) nam
ana

4.1.3 Philological Commentary

Here, and in the philological commentaries to the other tablets, a table with the divergent readings will show the parts of the transliteration where my proposal is different from those of other authors. Table 4.1, specifically, compares my proposed transliteration of IM55357 with those published by Baqir (1950a), Drenckhahn (1952a) and Høyrup (2002). In it, for instance, I state that kam, in line 3, is read kam$_2$ by Baqir. The reader is then allowed to conclude that both Drenckhahn and Høyrup read kam here. The same principle is followed throughout Table 4.1 and the tables corresponding to the other tablets.

Furthermore, there are a number of characteristics of the text that should be noticed:

- The several occurrences of igi.du$_3$ represent an unorthographic writing of igi.du$_8$.
- In line 8, the interrogative pronoun is rendered in this phrase in the accusative *mīnam*, following the reading of IM54478, where one finds "*mi-na-am* ib$_2$.si$_8$".
- In line 11, in order not to have 27 between a status constructus plus genitive construction, maybe the "grammatically correct" form should be 27 uš sag.du$_3$ an.ta.
- In line 16, I read a syllabic *ba*. It could be understood as the status constructus of a hypothetical **bûm* (*ba'um*). However, there are reasons to suppose that it comes from *bamtum* accompanied by a possessive suffix *-šu*, that, instead of the expected *bamassu*, produced the irregular **bamssum*, which was then abbreviated to *ba* (Høyrup 2002, 31, note 53).

Table 4.1 Divergent editorial readings in IM55357

Line	Sign(s)	Divergences
3	uš	Read TA by Baqir (1950a) and ta by Høyrup (2002)
3	kam	Read kam$_2$ by Baqir (1950a)
5	murgu$_2$	Read LUM by Baqir (1950a) and by Drenckhahn (1952a), as a logogram for *g/kaṣāṣum*
6	ta	Read da by Drenckhahn (1952a)
6	il$_2$	Always read ila$_2$ by Drenckhahn (1952a)
9	sag (erasure)	Read sag.du$_3$ (?)-x by Baqir (1950a)
10	gaba	Read TUḪ by Baqir (1950a)
12	tag$_4$	Read RU by Baqir (1950a)
E15	sag.ki <sag.>du$_3$	Read sag.ki du$_3$ by Baqir (1950a) and by Drenckhahn (1952a)
E15	kam$_2$	Read kam by Drenckhahn (1952a)
E16	*ba*	Read ba by Baqir (1950a) and Drenckhahn (1952a), BA by Høyrup (2002)
E16	*<ḫi>-pe$_2$*	Read bi (?) by Baqir (1950a)

As regards the expression za.e ki_3.ta.zu.un.ne, I follow Attinger (2005, 62) and consider it a variant writing of the standard expression za.e ak.ta.zu.ne (see Sect. 3.2). The introduction of the sign "un" may simply indicate that the scribe was trying to reproduce the well-known sound of a familiar expression but not entirely sure of the signs that should be used. For a different reading and translation, see Høyrup (2002, 32–33).

As a last observation, this text is unique among the twelve examined here in that it does not use the enclitic particle *-ma* to join sentences. This characteristic might be associated to its place in a different stratigraphic level.

4.1.4 Translation

[1]A triangle. 1,0 is the length, 1,15 is the long length, 45 is the upper width, [2]22,30 is the complete area. In 22,30, the complete area, 8,6 is the upper area, [3]5,11;2,24 is the {leaning/following} area, 3,19;3,56,9,36 is the third area, [4]5,53;53,39,50,24 is the lower area. [5]What are the upper length, the {length of the shoulder/rear length}, the lower length and the descendant? [6]You, in your doing, detach the *igi* of 1,0, the length. Raise to 45. [7,8]You see 0;45. Raise 0;45 to 2. You see 1;30. Raise 1;30 to 8,6, the upper area. You see 12,9. What does 12,9 make equal? It makes 27 equal. [9]27 is the width. Halve 27. You see 13;30. Detach the *igi* of 13;30. [10]Raise to 8,6, the upper area. You see 36, the length (which is) pair of the length 45, the width. [11]Return. Remove the length 27 of the upper triangle from 1,15. [12]48 {remains/is left behind}. Detach the *igi* of 48. You see 0;1,15. Raise 0;1,15 to 36. [13]You see 0;45. Raise 0;45 to 2. You see 1;30. Raise 1;30 to 5,11;2,24. [14]You see 7,46;33,36. What does 7,46;33,36 make equal? [E15]It makes 21;36 equal. 21;36 is the width of the second triangle. [E16]Halve the half of 21;36. You see 10;48. Detach the *igi* of 10;48. [E17]To...

4.1.5 Mathematical Commentary

The problem deals with a triangle of sides 1,0 (referred to as length), 1,15 (long length) and 45 (width). All this comes in line 1. Next, it is redundantly said that the area of the triangle is 22,30. The drawing that accompanies the text (as in Fig. 4.1, to which I added auxiliary letters) shows the triangle divided into four smaller triangles, with areas equal to

8,6
5,11;2,24
3,19;3,56,9,36
5,53;53,39,50,24

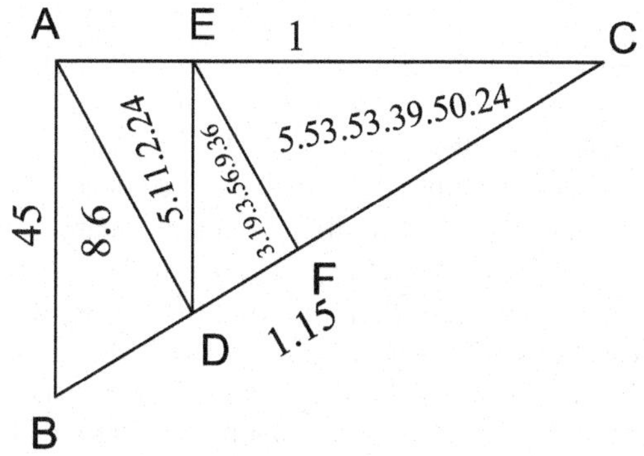

Fig. 4.1 The triangle in IM55357

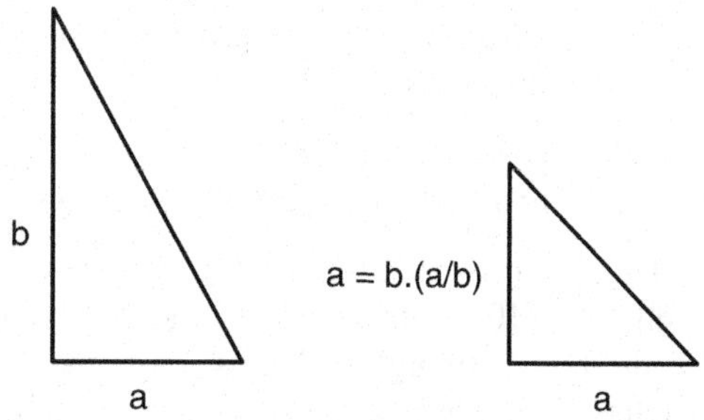

Fig. 4.2 Scaling a triangle to an isosceles triangle

These numbers are confirmed by the text of the problem, which furthermore gives them names: upper area, leaning/following area, third area and lower area (lines 2–4).

From these numbers, we are then able to conclude that these five triangles are (in our own terms) right triangles, and (again in our terms) they are similar to each other.

The problem asks for the lengths of four segments: the upper length, the length of the shoulder/the rear length, the lower length and the descendant (line 5). However, the text is not clear about the segments these expressions refer to. It is likely that the first three are intended to be the lengths of the first three smaller triangles, whereas the descendant could be either the long length or the length of the fourth triangle.

Accordingly, the required segments seem to be the upper length, AD; the length of the shoulder/the rear length, DE; the lower length, EF and the descendant, EC or CF.

In lines 6–10, the scribe makes a series of computations in order to obtain AD=36 (line 10: "You see 36, the length…"). This is accomplished through the application of the following two facts:

(I) If you double a right triangle, you obtain a rectangle.
(II) A right triangle with legs a and b can be transformed into an isosceles right triangle by means of a scaling of ratio a/b, in the direction of the leg of length b, as exemplified in Fig. 4.2. The area of the new triangle equals the area of the original triangle raised to a/b.

Thus, in the solution of the problem, the scribe initially obtains the *igi* of 1,0 and then raises it to 45 (line 6), arriving at 0;45, the ratio of the scaling that transforms triangle ABC into an isosceles right triangle of legs 45. After that, the ratio 0;45 is doubled, giving 1;30 (line 7). This number incorporates the ratio of the scaling as well as the doubling necessary to transform a right triangle into a rectangle, in this case an isosceles right triangle into a square.

In the next step, the scribe applies this ratio to the smaller triangle ABD. The scribe seems to tacitly employ the fact that the ratio that turns ABC into a square also turns ABD into a square. After raising the area 8,6 of triangle ABD to 1;30 (lines 7 and 8), the scribe obtains 12,9, which is the area of a square with sides BD (line 8). The scribe then asks for the value of the side of this square: "What does 12,9 make equal? It makes 27 equal. 27 is the width". (Lines 8 and 9.) In other words, BD=27. It is now time to go back to triangle ABD. As its area is 8,6 and as it is known that BD=27, the scribe is then able to obtain AD=36 by means of the following computations:

- Half of BD=13;30 (line 9).
- The *igi* of 13;30 raised to 8,6 is 36, the required value of AD (lines 9 and 10).

This result is presented as "36, the length (which is) pair of the length 45, the width" (line 10), that is to say, 36, the length which is pair of the length 45, the width (of the bigger triangle). This is, most likely, the upper length that the problem requires one to find.

DC is then calculated as 1,15−27=48 (line 11), and what follows from line 12 on is a repetition of the whole procedure:

- The *igi* of 48 is 0;1,15 (line 12).
- The raising of 0;1,15 to 36 is 0;45 (the ratio of the scaling that transforms ADC into an isosceles right triangle of legs 36) (lines 12 and 13).
- The raising of 0;45 to 2 is 1;30 (a number that also incorporates the doubling necessary to transform the isosceles right triangle into a square) (line 13).
- The raising of 1;30 to 5,11;2,24 is 7,46;33,36 (application of the ratio 1;30 to transform triangle AED into a square with sides AE) (lines 13 and 14).
- 7,46;33,36 makes 21;36 equal (this is the value of segment AE) (lines 14 and 15).

Thus, AE=21;36 is the "width of the second triangle" (line E15). In the last two lines, the scribe halves 21;36 and asks for its *igi*. The text of the tablet ends at this point, but it is clear that the raising of the *igi* of half of 21;36 to the area of AED would produce the length of the segment DE. Thus, if one completes what is lacking, one gets:

$$\begin{aligned}
&DE = \\
&\textit{igi}\text{ of the half of } 21;36 \text{ raised to } 5,11;2,24 = \\
&\textit{igi}\text{ of } 10;48 \text{ raised to } 5,11;2,24 = \\
&0;5,33,20 \text{ raised to } 5,11;2,24 = 28;48
\end{aligned}$$

Supposedly, DE is the "length of the shoulder", required by the statement of the problem.

As already told, the text of the solution ends before DE is obtained. However, the repetition of the above procedure would additionally enable one to obtain the lower length EF. It would then be enough to identify the ratio that transforms the right triangle EDC into a square of sides DE with the ratio that transforms the right triangle EDF into a square of sides DF.

Finally, the descendant required by the problem would be obtained simply by computing either the difference AC − AE = EC or BC − BF = CF.

4.1.6 Orders of Magnitude

The above translation and mathematical commentary present one possible interpretation of the orders of magnitude of the numbers that this tablet brings. However, other orders of magnitudes could have been chosen, without affecting the correctness of the mathematics involved. In the following, I copy the translation of the problem with a set of corresponding alternative values, which the interested reader will be able to easily check.

> [1]A triangle. 1;0 is the length, 1;15 is the long length, 0;45 is the upper width, [2]0;22,30 is the complete area. In 0;22,30, the complete area, 0;8,6 is the upper area, [3]0;5,11,2,24 is the {leaning/following} area, 0;3,19,3,56,9,36 is the third area, [4]0;5,53,53,39,50,24 is the lower area. [5]What are the upper length, the {length of the shoulder/rear length}, the lower length and the descendant? [6]You, in your doing, detach the *igi* of 1;0, the length. Raise to 0;45. [7,8]You see 0;45. Raise 0;45 to 2. You see 1;30. Raise 1;30 to 0;8,6, the upper area. You see 0;12,9. What does 0;12,9 make equal? It makes 0;27 equal. [9]0;27 is the width. Halve 0;27. You see 0;13,30. Detach the *igi* of 0;13,30. [10]Raise to 0;8,6, the upper area. You see 0;36, the length (which is) pair of the length 0;45, the width. [11]Return. Remove the length 0;27 of the upper triangle from 1;15. [12]0;48 {remains/is left behind}. Detach the *igi* of 0;48. You see 1;15. Raise 1;15 to 0;36. [13]You see 0;45. Raise 0;45 to 2. You see 1;30. Raise 1;30 to 0;5,11,2,24. [14]You see 0;7,46,33,36. What does 0;7,46,33,36 make equal? [E15]It makes 0;21,36 equal. 0;21,36 is the width of the second triangle. [E16]Halve the half of 0;21,36. You see 0;10,48. Detach the igi of 0;10,48. [E17]To …

So we see that, due to the absence of the integral–fractional separator (the semicolon used in the translations and commentaries), cuneiform mathematical texts can be interpreted by historians of mathematics in more than one way. This should not be seen as a proof that scribes were making all their arithmetic operations under some organised concept of equivalence classes, but the fact helps us increase our historiographic sensitiveness, for it shows that the reading of numbers in cuneiform tablets fits a different cognitive numerical formation.[2] An illuminating comparison

---

[2]For the benefit of the mathematically minded reader, it should be said that these equivalence classes would not be the usual ones obtained in the modulus-60 arithmetics. Instead, two numbers are equivalent here if one is the other multiplied by a power of 60. So it is not true that 62 and 2 are equivalent in this floating point system. But 2,0 (that is to say, 120) and 2 are indeed equivalent (Proust 2013).

between the floating point arithmetic in cuneiform mathematics and the use of the slide rule in engineering courses until the 1950s is made by Høyrup (2012): in neither of the contexts, there is a need for written integral–fractional separators, but both the scribe and the engineer are, in principle, able to keep record of the orders of magnitudes of the numbers they manipulate.

4.2 IM52301

4.2.1 References, Physical Characteristics and Contents

The obverse and the reverse are completely occupied by two problems followed by a list of coefficients. As usual, the right edge receives the ending of the lines that cannot be written entirely on the two faces of the tablet. In the left edge, in addition, there is a so-called third problem, which consists of a small text, of rather difficult interpretation, which might be a general rule, carrying the main ideas of the problems on the faces of the tablet.

Baqir (1950b) wrote the original publication. Bruins (1951) presented an alternative mathematical interpretation for the second problem (and a new reading for one sign in line 17), stating that it deals with a divided triangle and not a trapezium, as Baqir had originally argued. With von Soden (1952), new readings for some very few passages were presented. Bruins (1953a) resumed this tablet (with two others) to contest a philological point of von Soden's publication and to offer an interpretation for the difficult Problem 3, stating *wrongly* that it is a shortened version of the formula for obtaining the square root of a number. A complete new transliteration of the tablet and a corresponding mathematical interpretation in terms of general quadrilaterals were the contents of Gundlach and von Soden (1963). Jens Høyrup (1990, 2002) concentrated on aspects of Problem 2. Robson (1999) completely transcribed and analysed the list of coefficients. Finally, Friberg (2000) explained Problem 3 as a general formulation for the so-called *Surveyor's Formula* for the area of a quadrilateral, an interpretation that is adopted here. In studying this tablet, I also had access to duplicated prints of the photos published by Baqir (1950b). These photos are most likely the ones used by Gundlach and von Soden (1963), as they indicate in their paper.[3]

The tablet comes from Level II of room 180 of Tell Harmal (Hussein 2009, 92). Its dimensions are 15.9 × 9.7 × 3.4 cm.

[3] Here too, I am indebted to Hermann Hunger for the access to the photos.

4.2.2 Transliteration and Transcription

Problem 1

Obverse

(1) *šum-ma* 1.40 uš *e-lu-um me-ḫe-er-šu ḫa-li-iq* sag.ki *e-li-tum*
šumma 1.40 *šiddum elûm meḫeršu ḫaliq pūtum elītum*
(2) *e-li* sag.ki *ša-ap-li-tim* 20 *e-te-er* 40 a.ša$_3$ *mi-nu-um* uš-*ia-ma*
eli pūtim šaplītim 20 *etter* 40 *eqlum mīnum šiddija-ma.*
(3) za.e kid$_2$(?).zu$_2$.ne 1.30 *šu-ku-un-ma ḫi-pe*$_2$ <<*šu-ta-ki-il*>>*-ma*
atta ina epēšika 1.30 *šukun-ma ḫipē-ma*
(4) 45 *ta-mar* igi 45 duḫ.ḫa-*ma* 1.20 *ta-m*[*ar* 1.20 *a*]-*na* 40 a.ša$_3$ *i-ši-ma*
45 *tammar. igi* 45 *puṭur-ma* 1.20 *tammar.* 1.20 *ana* 40 *eqlim išī-ma*
(5) 53.20 *ta-mar* 53.20 *e-ṣi*$_2$-ma 1.46.40 [*ta*]-*mar* 1.46.40 *re-eš*$_{15}$*-ka*
53.20 *tammar.* 53.20 *eṣim-ma* 1.46.40 *tammar.* 1.46.40 *rēška*
(6) *li-ki-il tu-ur-ma* 1.40 uš *a-li-a-am u*$_3$ 20 *ša* sag.ki *e-li-tum*
likīl. tūr-ma. 1.40 *šiddam aliam u* 20 *ša pūtum elītum*
(7) *e-li* sag.ki *ša-ap-li-tim i-te-ru ku-mu-ur-ma* 2 *ta-mar*
eli pūtim šaplītim itteru kumur-ma 2 *tammar.*
(8) 2 *ḫi-pe*$_2$*-ma šu-ta-ki-il-ma* 1 *ta-mar* 1 *a-na* 1.46.40 *ṣi*$_2$*-ib-ma*
2 *ḫipē-ma. šutākil-ma* 1 *tammar.* 1 *ana* 1.46.40 *ṣib-ma*
(9) 2.46.40 *ta-mar ba-se-e* 2.46.40 *šu-li-*[*ma*] 1.40 *ta-mar*
2.46.40 *tammar. basê* 2.46.40 *šūli-ma* 1.40 *tammar.*
(10) *a-na* 1.40 *ba-si-ka* 1 *ša tu-uš-ta-ki-lu* <<*a-na* 1.40>> *ṣi*$_2$-[*ib*]*-ma*
ana 1.40 *basika* 1 *ša tuštākilu ṣib-ma*
(11) 2.40 *ta-mar i-na* 2.40 *ša ta-mu-ru* 1.40 uš *a-li-am ḫu-ru-uṣ*$_4$
2.40 *tammar. ina* 2.40 *ša tamuru* 1.40 *šiddam aliam ḫuruṣ.*
(12) 1 *ši-ta-tum* uš *ḫa-al-qu*$_2$ 1 *ḫi-pe*$_2$*-ma* 30 *ta-mar* 30 *me-eḫ-ra-am*
1 *šittātum šiddum ḫalqu.* 1 *ḫipē-ma* 30 *tammar.* 30 *meḫram*
(13) *i-di-ma* 20 *ša* sag.ki *e-li* sag.ki *i-te-ru ḫi-pe*$_2$*-ma*
idī-ma. 20 *ša pūtum eli pūtim itteru ḫipē-ma*
(14) 10 *ta-mar* 10 *a-na* 30 *iš-ten ṣi*$_2$*-ib-ma* 40 *ta-mar i-na* 30 *ša-ni-im*
10 *tammar.* 10 *ana* 30 *ištēn ṣib-ma* 40 *tammar. ina* 30 *šanîm*
(15) *ḫu-ru-uṣ*$_4$ 20 *ta-mar* 20 sag.ki *ša-ap-li-tum ki-a-am ne-pe*$_2$*-šum*
ḫuruṣ. 20 *tammar.* 20 *pūtum šaplītum. kīam nēpešum*

Problem 2

(16) *šum-ma a-na ši-ni-ip ku-mu-ri* sag.ki *e-li-tim*
šumma ana šinīp kumurrî pūtim elītim
(17) *u*$_3$ *ša-ap-li-tim* 10 *a-na qa-ti-ia* taḫ$_2$*-ma* {20/min$_3$} uš *ab-ni* sag.ki
u šaplītim 10 *ana qatija ūṣib-ma* {20/*šaniam*} *šiddam abni pūtum*

(18) *<<e-li>> e-li-tum e-li ša-ap-li-tim* 5 *i-te-er*
elītum eli šaplītim 5 *itter*
(19) a.ša$_3$ 2.30 *mi-nu* uš-*ia* za.e kid$_2$(?).zu$_2$.ne 5 *ša e-te-ru*
eqlum 2.30 *mīnu šiddija. atta ina epēšika* 5 *ša etteru*
(20) 10 *ša tu-iṣ-bu* 40 *ši-ni-pe*$_2$*-tim a-ra-ma-ni-a-ti-a lu-pu-ut-ma*
10 *ša tuṣbu* 40 *šinipêtim aramaniātija luput-ma.*
(21) *i-gi* 40 *ši-ni-pe*$_2$*-tim pu-ṭu*$_2$*-ur-ma* 1.30 *ta-mar* 1.30 *<<ḫi-pe*$_2$!(*du*$^?$)*-ma*
igi 40 *šinipêtim puṭur-ma* 1.30 *tammar.* 1.30 *ḫipē-ma.*
(22) 4[5 *t*]*a-mar* 45>> *a-na* 2.30 a.ša$_3$ *i-ši-ma* 3.45 *ta-mar*
45 *tammar.* 45 *ana* 2.30 *eqlim išī-ma* 3.45 *tammar.*
(23) 3.45 *e-ṣi*$_2$*-ma* 7.30 *ta-mar* 7.30 *re-eš*$_{15}$*-ka*
3.45 *eṣim-ma* 7.30 *tammar.* 7.30 *rēška*
(24) *li-ki-il tu-ur-ma i-gi* 40 *ši-ni-pe*$_2$*-tim pu-ṭu*$_2$*-ur*
likīl. tūr-ma. igi 40 *šinipêtim puṭur.*

Reverse

(R1) 1.30 *ta-mar* 1.30 *ḫi-pe*$_2$*-ma* 45 *ta-mar a-na* 10 *ša tu-iṣ-bu*
1.30 *tammar.* 1.30 *ḫipē-ma* 45 *tammar. ana* 10 *ša tuiṣbu*
(R2) *i-ši-ma* 7.30 *ta-mar* <<7.30 *re-eš*$_{15}$*-ka li-ki-il*
išī-ma 7.30 *tammar.* 7.30 *rēška likīl.*
(R3) *tu-ur-ma i-gi* 40 *pu-ṭu*$_2$*-ur-ma* 1.30 *ta-mar* 1.30!(40) *ḫi-pe*$_2$*-ma*
tūr-ma igi 40 *puṭur-ma* 1.30 *tammar.* 1.30 *ḫipē-ma.*
(R4) 45 *ta-mar a-na* 10 *ša tu-iṣ-bu i-ši-ma* 7.30 *ta-mar*>>
45 *tammar. ana* 10 *ša tuiṣbu išī-ma* 7.30 *tammar.*
(R5) 7.30 *me-eḫ-<<ša>>-ra-am i-di-ma šu-ta-ku-il-ma*
7.30 *meḫram idī-ma šutākil-ma.*
(R6) 56.15 *ta-mar* 56.15 *a-na* 7.30 *ša re-eš*$_{15}$*-ka*
56.15 *tammar.* 56.15 *ana* 7.30 *ša rēška*
(R7) *u*$_2$*-ka-lu ṣi*$_2$*-ib-ma* 8.26.15 *ta-mar ba-se-e*
ukallu ṣib-ma 8.26.15 *tammar. basê*
(R8) 8.26.15 *šu-li-ma* 22.30 *ba-su-šu i-na* 22.30
8.26.15 *šūli-ma.* 22.30 *basûšu. ina* 22.30
(R9) *ba-se-e* 7.30 *ta-ki-il-ta-ka ḫu-ru-uṣ*$_4$
basê 7.30 *takīltaka ḫuruṣ.*
(R10) 15 *ši-ta-tum* 15 *ḫi-pe*$_2$*-ma* 7.30 *ta-mar* 7.30 *me-eḫ-ra-am i-di-ma*
15 *šittātum.* 15 *ḫipē-ma* 7.30 *tammar.* 7.30 *meḫram idī-ma.*
(R11) 5 *ša* sag.ki *e-li* sag.ki *i-te-ru ḫi-pe*$_2$*-ma*
5 *ša pūtum eli pūtim itteru ḫipē-ma.*
(R12) 2.30 *ta-mar* 2.30 *a-na* 7.30 *iš-ti-in ṣi*$_2$*-im-ma*
2.30 *tammar.* 2.30 *ana* 7.30 *ištīn ṣim-ma*
(R13) 10 *ta-mar i-na* 7.30 *ša-ni-im ḫu-ru-uṣ*$_4$
10 *tammar. ina* 7.30 *šanîm ḫuruṣ.*
(R14) 10 sag.ki *e-li-tum* 5 sag.ki *ša-ap-li-tum*
10 *pūtum elītum.* 5 *pūtum šaplītum.*
(R15) *tu-ur-ma* 10 *u*$_3$ 5 *ku-mu-ur* 15 *ta-mar*
tūr-ma. 10 *u* 5 *kumur.* 15 *tammar.*

(R16) *ši-ni-ip-pe$_2$-at* 15 *le-qe$_2$-ma* 10 *ta-mar u$_3$* 10 *ṣi$_2$-ib-ma*
šinipeat 15 *leqē-ma.* 10 *tammar u* 10 *ṣib-ma.*
(R17) 20 uš-*ka e-lu-um* 15 *ḫi-pe$_2$-ma* 7.30 *ta-mar*
20 *šidduka elûm.* 15 *ḫipē-ma* 7.30 *tammar.*
(R18) 7.30 *a-na* 20 *i-ši-ma* 2.30 a.ša$_3$ *ta-mar*
7.30 *ana* 20 *išī-ma* 2.30 *eqlam tammar.*
(R19) *ki-a-am ne-pe$_2$-šum*
kīam nēpešum.

List of Coefficients

(R20) 6.40 *i-gi-gu-ub-bi-im qu$_2$-up-pi$_2$-im*
6.40 *igigubbîm quppim.*
(col I - right)
(R21) 3.45 *pi$_2$-ti-iq-tum*
3.45 *pitiqtum.*
(R22) 4.10 *ša* šeg$_{12}$
4.10 *ša {libitti/libittī}.*
(R23) 5 *ša ki-pa-ti*
5 *ša kippati.*
(R24) 30 *ša sa$^?$-ta$^?$-[ki]$^?$*
30 *ša sattakki.*
(Col II - left)
(R21) 6 *ša na-aš-pa-kum*
6 *ša našpakum.*
(R22) 7.30 *ka-ru-um*
7.30 *karûm.*

Problem 3[4]

Edge

(E1) *šum-ma* a.ša$_3$ uš *la mi-it-ḫa-ru-ti at-ta // i-gi* 4 *pu-ṭu$_2$-ur-ma*
šumma eqel šiddī lā mitḫārūti, atta // igi 4 *puṭur-ma.*
(E2) *na*!*-ap-ḫa-ar* uš *li-iq-bu-ni-kum-ma // a-na na-ap-ḫa-<ar> ši-di-ka i-ši-ma*
napḫar šiddī liqbûnikkum-ma. // ana napḫar šiddīka išī-ma.
(E3) 4$^?$ *ša-ar er-be$_2$-tim lu-<pu>-ut-ma // ma*!*-la*! *i-li-ku tu-uš-ta-ka-al-ma i-na li-ib-bi-<šu$^?$>*
4 *šār erbettim luput-ma. // mala illiku tuštākal-ma. // ina libbišu*
(E4) a.ša$_3$ *ta-na-sa-aḫ.*
eqlam tanassaḫ.

[4] The first three lines on the edge seem to have been written in two columns. This produces, for each line, an initial and a final segment. The separation is marked by double slashes.

4.2.3 Philological Commentary

In what follows, Table 4.2 summarises the divergences between my transliterations and those by Baqir (1950b), Gundlach and von Soden (1963), Høyrup (1990), Friberg (2000) and Høyrup (2002). In reading the table, one must take into account that both Høyrup (1990) and Høyrup (2002) offered only the transliteration of Problem 2 (lines 16 to R19), whereas Gundlach and von Soden (1963) transliterated the whole of the tablet, except the coefficient list. On the other side, Friberg (2000) gives only the transliteration of Problem 3. It must also be said that the coefficient list is not covered by the table. Finally, the table lists only differences in the identification of signs: divergences like *ri-iš-ka* (Baqir (1950b), Høyrup (1990), Gundlach and von Soden (1963)) and *re-eš*$_{15}$*-ka* (Høyrup (2002) and the present work) have not been registered.

Apart from the divergences exposed in the previous table, there is a sign in line 17 that causes some difficulty too. MAN is read *killalan* by Baqir and 20 by Høyrup, according to whom 20 "is meant for naming, not as a datum" (2002, 215). Bruins (1951) read it as *šanûm*, "other, second", which I believe is the reading intended by the scribe. I transliterate it as min$_3$. It might be useful to know that *šanûm* also appears in IM54478, with the same meaning (although syllabically written). It appears syllabically written in line 14 too.

The text deviates from the other analysed tablets in the writing of the word for the number one. While all others write *iš-te-en*, IM52301 writes *iš-ti-in* or *iš-ten*.

Table 4.2 Divergent editorial readings in IM52301

Lines	Signs	Divergences
4	duḫ	Read tuḫ by Baqir (1950b)
11, 15, R9, R13	*ḫu-ru-uṣ*$_4$	Read *ḫu-ru-uṣ*$_2$ by Baqir (1950b)
12	*ḫa-al-qu*$_2$	Read *ḫa-al-qu* by Baqir (1950b)
14	*iš-ten ṣi*$_2$*-ib-ma*	Read *iš-te-en ṣi*$_2$*-ib-ma* by Gundlach and von Soden (1963)
17	taḫ$_2$	Read taḫ by Baqir (1950b), daḫ by Høyrup (1990, 2002), DAḪ by Gundlach and von Soden (1963)
17	20/min$_3$	Read *kilallâ* by Baqir (1950b), 20 by Høyrup (1990, 2002) and Gundlach and von Soden (1963)
19	*mi-nu*	Read *mi-nu-um* by Baqir (1950b), Høyrup (1990, 2002) and Gundlach and von Soden (1963)
21	*ḫi-pe*$_2$!(*du*?)	Read *ḫe-pe*$_2$(?) by Baqir (1950b), *ḫi-pi*$_2$(?) by Høyrup (1990), *ḫe-*¿*pe*$_2$$^?$ by Høyrup (2002), *ḫe-pe*$_2$(?) by Gundlach and von Soden (1963)
E2	<*ar*> *ši-di-ka*	Read *ar* sag by Friberg (2000)
E4	*ta-na-sa-aḫ*	Read *ta-tam-ša*$_{10}$*-aḫ* by Friberg (2000)

In lines 2 and 19, we have the first person *e-te-er* and *e-te-ru*. The occurrence in line 2 leads to a difficult syntactical arrangement. One may ask whether it was intended to be a third person singular. If so, one could ask additionally whether this instability e/i reveals something about the pronunciation of the initial vowel as an intermediary sound between e and i.

In line 3, I read <<*šu-ta-ki-il*>> as a scribal copy mistake. This might have been caused by an interference of the sequence 2 *ḫi-pe*$_2$*-ma šu-ta-ki-il-ma*, where 2 is really halved and the result 1 is squared. In line 3, however, 1.30 is halved, producing 45, but nothing is squared.

Also in line 3, *šu-ku-un-ma* might be pronounced *šukum-ma*, with the final n of the verb assimilated to m (See also IM54538, lines 5 and 5; GAG §33h).

Some further observations are:

- In lines 3 and 19, we find what seems to be a different writing of the Sumerian expression za.e ak.ta.zu.ne. This difference offers room for some speculation. Baqir (1950b) transliterated the first sign as tug, leading to a reading equivalent to tug.zu$_2$.ne. He also suggested that tug might be the verb "to say" (Sumerian dug, normally written dug$_4$, but here presented unorthographically[5] as tug). Baqir's reading must be taken seriously, because he had the tablet on his hands and, perhaps anticipating that this would lead to some discussion, were careful enough to register that the signs were clearly written (Baqir 1950b, 146). However, I would like to suggest another possibility, namely, to read the first sign as kid$_2$. Although tug and kid$_2$ are different signs, some of their written versions show a certain degree of similarity. Thus, for some reason, the scribe might have been trying to reproduce the familiar sound of the expression, but using signs that were not the traditional ones: kid$_2$ instead of the expected ak (the same sign as kid$_3$) and zu$_2$ instead of the expected zu. In the same line of reasoning, the absence of the sign "ta" may simply indicate that the scribe guaranteed the dental sound "t/d" at the end of kid$_2$ and felt free to omit the intermediary "a". That phonetic issues might have been acting here is consistent with the use of the same expression in the previous tablet, IM55357, where the addition of the sign "un" might have been due only to the excess of zeal of the scribe to guarantee its presence in pronunciation.
- In line 5, we see a not so common assimilation of p to m: *eṣip-ma* > *eṣim-ma* (G-Stem Imp. 2. Sg. m.). On the other hand, the assimilation *ṣi*$_2$*-ib-ma* > *ṣi*$_2$*-im-ma* (reverse, line 12) is a common and expected one, once b is a voiced consonant (GAG §27c).
- In line 19, Gundlach and von Soden transcribe *šiddū-ia*.
- In line 20, *aramaniāti-ja* is an untranslated term. The only example in both CAD and AHw is the present one. von Soden (1952, 50) conjectures that it is a loanword

[5] The concept of unorthographic writing serves to explain deviations from expected writings. In the present example, dug$_4$ is the expected way a scribe writes the verb to say, but as the phonetic values of the sign tug include the pronunciation "dug", this sign can be used instead, constituting an unexpected or unorthographic writing.

from Sumerian ara.man, double factor, indicating the multiplier of a sum. However, as he argued, further attestations should be necessary to elucidate the question definitely.

- In lines 21 and 22, Gundlach and von Soden do not delete anything, but they propose the insertion … 45 <*eṣip-ma* 1,30 *ta-mar* 1,30> *a-na* …
- In line R7, we see the subjunctive *ukallu*, with the same phonetic changes as in plural forms.
- In line R21 Col II, a genitive was expected here: *ša pi*$_2$*-ti-iq-tim*, instead of the written *pi*$_2$*-ti-iq-tum.*
- In line E4, we read *eqlim* in von Soden (1952).

4.2.4 *Translation*

Problem 1

[1]If 1,40 is the upper length, (if) its opposite is missing, (if) the upper width: [2]I go 20 beyond the lower width, (if) 40,0 is the area, what is my length? [3]You, in your doing, place 1;30 and halve it, and [4]you see 0;45. Detach the *igi* of 0;45, and you see 1;20. Raise 1;20 to 40,0, the area, and [5,6,7]you see 53,20. Double 53,20, and you see 1,46,40. May 1,46,40 hold your head. Return. Accumulate 1,40, the upper length, and 20 that the upper width goes beyond the lower width, and you see 2,0. [8]Halve 2,0. Combine (the halves), and you see 1,0,0. Add 1,0,0 to 1,46,40, and [9]you see 2,46,40. Cause to come up the equal of 2,46,40, and you see 1,40. [10]Add to 1,40, your equal, 1,0 that you have combined, and [11]you see 2,40. Cut off 1,40, the upper length, from 2,40 that you saw. [12,13]1,0, the remainder, is the missing length. Halve 1,0, and you see 30. Write down 30, the copy. Halve 20 that the width goes beyond the width, and [14,15]you see 10. Add 10 to the first 30, and you see 40. Cut off from the second 30. You see 20. The lower width is 20. Thus, the procedure.

Problem 2

[16,17,18]If I added 10, on my hand, to the two thirds of the accumulation of the upper and the lower width, I built {a second length/20, the length}, (if) the upper width goes 5 beyond the lower (width), [19,20](if) the area is 2,30, what is my length? You, in your doing, record 5 that I go (beyond), [20]10 that you added, 0;40, the two thirds, my *aramanitum*. [21,22]Detach the *igi* of 0;40, the two thirds, and you see 1;30. Raise 1;30 <<halve and you see 0;45. 0;45>> to 2,30, the area, and you see 3,45. [23,24]Double 3,45, and you see 7,30. May 7,30 hold your head. Return and detach the *igi* of 0;40, the two thirds. [R1,R2]You see 1;30. Halve 1;30, and you see 0;45. Raise to 10 that you added, and you see 7;30. <<May 7;30 hold your head. [R3]Return. Detach the *igi* of 0;40, and you see 1;30. Halve 1;30, and [R4]you see 0;45. Raise to 10 that you added,

and you see 7;30.>> [R5]Write down 7;30, the copy, and combine (them). [R6,R7,R8,R9]You see 56;15. Add 56;15 to 7,30 that holds your head, and you see 8,26;15. Cause to come up the equal of 8,26;15, and its equal is 22;30. Cut off 7;30, your *takiltum*, from 22;30, the equal. [R10]The remainder is 15. Halve 15, and you see 7;30. Write down 7;30, the copy. [R11]Halve 5 that the width goes beyond the width, and [R12]you see 2;30. Add 2;30 to the first 7;30, and [R13]you see 10. Cut off from the second 7;30. [R14]The upper width is 10. The lower width is 5. [R15]Return. Accumulate 10 and 5. You see 15. [R16]Take the two thirds of 15. You see 10 and add 10. [R17]Your upper length is 20. Halve 15, and you see 7;30. [R18]Raise 7;30 to 20 and you see 2,30, the area. [R19]Thus, the procedure.

List of Coefficients

6,40,0 of the coefficient (of) a basket. 0;3,45 a brickwork. 0;4,10 of {a brick/bricks}. 0;5 of a circle. 0;30 of a triangle. 6,0,0[6] of a storehouse. 7,30,0 the pile of barley.

Problem 3[7]

[E1]If an area is (made) of lengths that are not equal to each other, you: [E2]they should say to you the sum of the lengths. [E3]Write down 4, the four directions. [E1//]Detach the *igi* of 4, and [E2//]raise to the sum of your lengths. [E3//]As much as it comes, you combine and, from its interior, [E4]you remove the area.[8]

4.2.5 Mathematical Interpretation of Problems 1 and 2

Introductory Remarks

The Given Data and the Questions

Problem 1 begins with an exposition of the given data: the upper length is 1,40; the difference between the upper and the lower width is 20 (the difference is taken in this order, that is to say, the upper width is bigger than the lower width) and the area is 40,0. Thus, the problem deals with a quadrilateral whose sides are named upper length, lower length, upper width and lower width.

[6] See mathematical commentary for the order of magnitude.

[7] Following Friberg's (2000, 118) interpretation.

[8] The second segments of lines E1, E2 and E3 on the edge are marked with double slashes: E1//, E2// and E3//.

The given data in Problem 2 is slightly harder to understand: by adding 10 to two thirds of the accumulation of the upper and lower width, a certain length is built; the difference between the upper and the lower width (still in this order) is 5 and the area is 2,30.

The statement of this problem has the interesting mention of a hand. This might be a metaphor to a computation device, some sort of reckoning board, of which however we do not know the exact details, as suggested by Proust (2000). It is tempting to assume that 10 was obtained previously in this device, but of course there is no material basis for believing this was really the case. Anyway, instead of 10, we could have another number, and the problem shows how to cope with this situation by means of a paradigmatic example.

In both problems, the question is "What is my length?"

Before we proceed reading the text of these problems, five preparatory observations must be made.

Observation 1: Which Quadrilaterals Are Dealt with in These Problems?

Firstly, while Problem 1 makes in its statement a clear distinction between an upper and a lower length, Problem 2 simply refers to a general length. This may be an indication that in Problem 2 the geometrical figure dealt with is defined by only one length. Thus, it is possibly a right trapezium, and this length is the distance between the parallel bases, referred to as widths (Høyrup 2002). This is the interpretation adopted here. An alternative interpretation was given by Gundlach and von Soden (1963), who proposed that the upper length and the lower length were equal. This might indeed be what the scribe meant, but fortunately enough this would not change drastically the discussion contained in the following paragraphs. So, whenever a trapezium is dealt with in what follows, the reader may reproduce a similar statement with a quadrilateral possessing equal lengths instead.

Observation 2: There Is Something Missing in Problem 1. The Need for Additional Data

The second remark comes from that in Problem 2 widths and lengths are so that by adding 10 to two thirds of the accumulation of the upper and lower widths, the length is built, while in Problem 1 there is not any piece of information like this, linking lengths to widths. In fact, this absence makes Problem 1 an undetermined one, in the sense that it is not possible to calculate the missing sides only from the data that was explicitly given. Furthermore, a particular passage in the text of Problem 1 seems to confirm that there was indeed additional data that became absent in the written version of the problem. It goes like this: "Accumulate 1,40, the upper length, and 20 that the upper width goes beyond the lower width, and you see 2,0" (lines 6 and 7). Without additional data, it is not possible to know the rationale of this addition (even though it is clear that it is numerically right). So, starting from these observations, both Gundlach and von Soden (1963) and Høyrup (without date) completed Problem 1 with the following piece of information:

Additional data—Version A: *I accumulated the upper and the lower length and I cut off the accumulation of the upper length and* 20, *that the upper width goes beyond the lower width. I raised to* 1;30, *and I built the accumulation of the upper and lower widths.*

This is in fact consistent with the widths and lengths of the quadrilateral, as we see in the solution of the problem: upper length 1,40; lower (originally missing) length 1,0; upper width 40 and lower width 20.

Numerically speaking, this additional statement corresponds to the following operations:

- I accumulated the upper and the lower lengths: 1,40 accumulated with 1,0 is 2,40.
- I cut off the accumulation of the upper length and 20, that is to say, I cut off the accumulation of 1,40 and 20. Thus, I cut off 2,0
- I have 2,40 from which I cut off 2,0. I have 40.
- I raise it to 1,30. It is 1,0. This is indeed the same as the accumulation of the upper and lower widths, that is to say, the accumulation of 40 and 20.

Of course, it is not possible for us to know if this is indeed what the scribe meant. Even in the positive case, we cannot afford to state that we know what the exact words of the scribe would have been. In order to illustrate this point, here are two other ways in which it could have been written by the scribe.

Additional data—Version B: *I accumulated the upper and lower lengths and I cut off the accumulation of the upper length and* 20, *that the upper width goes beyond the lower width. I built two thirds of the accumulation of the upper and the lower widths.*

Numerically, we have:

- I accumulated the upper and the lower lengths: this is the accumulation of 1,40 and 1,0, which is 2,40.
- I cut off 2,0, which is the accumulation of the upper length 1,40 and 20. I get 40. This is indeed two thirds of the accumulation of 40 and 20 (the accumulation of the upper and the lower widths).

Finally,

Additional data—Version C: *To two thirds of the accumulation of the upper and the lower widths, I added the accumulation of the upper length and* 20, *that the upper width goes beyond the lower width. I build the accumulation of the upper and lower lengths.*

Numerically, we have:

- Two thirds of the accumulation of the upper (40) and the lower (20) widths equal two thirds of 1,0, that is to say, 40. I added 2,0, which is the accumulation of the upper length (1,40) and 20. I get 2,40. This is indeed the accumulation of the upper and the lower lengths.

Observation 3: The Accumulation of the Upper Length and 20, that the Upper Width Goes Beyond the Lower Is 2,0

This accumulation will appear quite frequently in what follows. Thus, it might be useful to remark from the beginning that its value is 2,0, once the upper length is 1,40. By the way, this value is explicitly calculated in lines 6 and 7, as already mentioned.

Observation 4: The Surveyor's Formula for the Area

In Problem 1, the area seems to be calculated by means of the well-known approximation procedure, known as surveyor's formula, stating that the area of a quadrilateral is obtained by raising the average width to the average length. In modern terms, by multiplying the average length and the average width. It is interesting to notice that this seems to be the procedure formulated in general terms in Problem 3.

Observation 5: Avoiding Symbolic Algebra

In order to analyse this problem without resorting to symbolic algebra, we may notice the following. Associated to a quadrilateral, it may be meaningful to speak of a direction of the lengths and a direction of the widths, as in Fig. 4.3.

Also, the approximate formula for the area—as the raising of half the accumulation of the lengths by half the accumulation of the widths—gives us a concrete way of seeing such accumulations: four copies of the quadrilateral, made along the two directions suggested above, cover a region that has these accumulations as approximate dimensions, as in Fig. 4.4. These approximations will hereafter be referred to simply as accumulation of the lengths and accumulation of the widths, in the spirit of the Old Babylonian formula for the area, never labelling these accumulations as approximations.

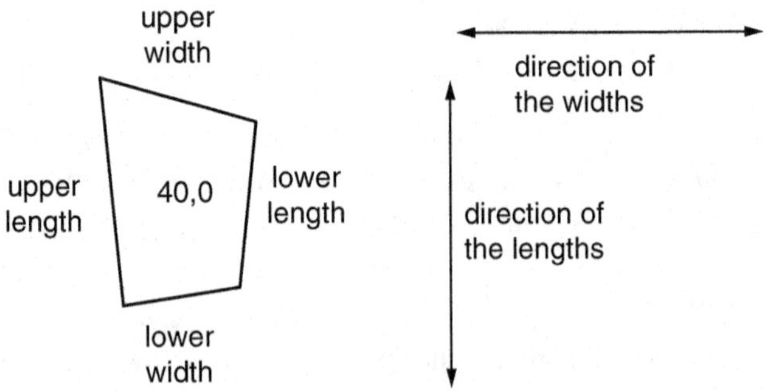

Fig. 4.3 Direction of the lengths and direction of the widths

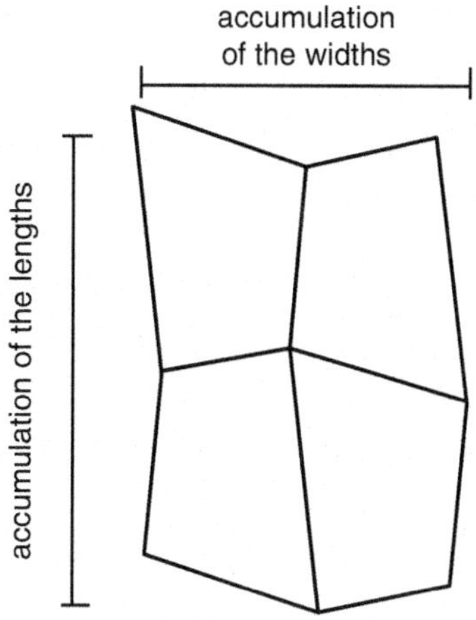

Fig. 4.4 Four copies of the quadrilateral, showing accumulation of the lengths and accumulation of the widths

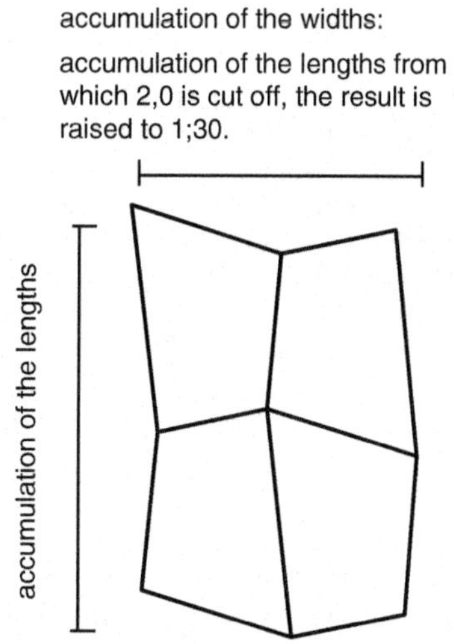

Fig. 4.5 Version A of the additional data. The accumulation of the widths expressed from the accumulation of the lengths

As a consequence, we have a very convenient way of expressing the additional data, in all its versions, as is shown in Figs. 4.5, 4.6 and 4.7. In Figs. 4.6 and 4.7, specifically, the quadrilateral has been compressed by a ratio of two thirds in the direction of the widths.

Now we are ready to read Problem 1.

Problem 1

The scribe starts solving the problem by placing the number 1;30 (line 3), which is present in version A of the additional data. He halves it (line 3), obtaining 0;45, and then he detaches the *igi* of 0;45, which is 1;20 (line 4). The area is raised to 1;20,

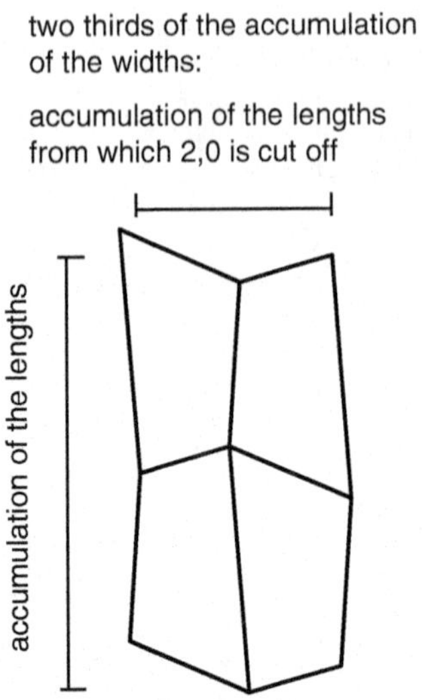

Fig. 4.6 Version B of the additional data. Two thirds of the accumulation of the widths expressed from the accumulation of the lengths

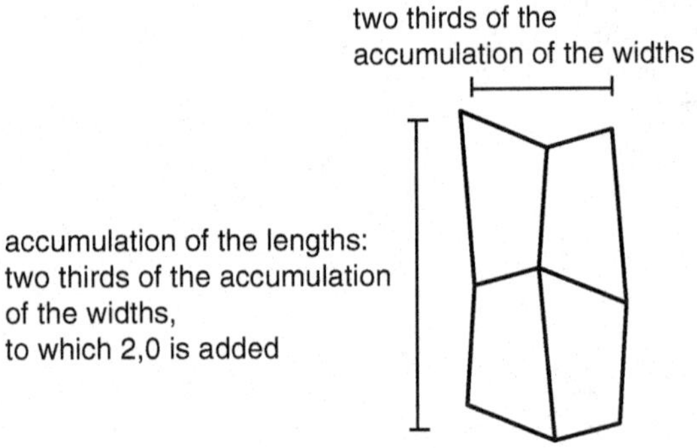

Fig. 4.7 Version C of the additional data. The accumulation of the lengths expressed from the two thirds of the accumulation of the widths

and the result is 53,20 (line 5), which is then doubled, giving 1,46,40 (line 5). Finally, this value must be left at our disposition (may it "hold your head", lines 5 and 6). All in all, these steps raise the original area of the quadrilateral to the *igi* of 1;30 (which is two thirds) to 2 and to 2 again. Consequently, these steps can be interpreted as the building of the configuration shown in Fig. 4.7, where we have four copies of a scaling of the original quadrilateral. This scaling was conveniently made along the direction of widths and has ratio two thirds (the *igi* of 1;30, as already commented). The rationale behind these operations is that the scribe now knows that (a) two thirds of the accumulation of the widths and (b) the accumulation of the lengths are two numbers for which the raising is 1,46,40 and of which the difference is 2,0 (as a consequence of the additional data). In other words, the problem is reduced to one about a square, as in Fig. 4.8a, where to an unknown square a rectangle of width 2,0 is added, producing a rectangle of area 1,46,40. Consistency with what I called additional data seems to be given by the subsequent lines of the text on the tablet, for in lines 6 and 7 the scribe accumulates the upper length (known

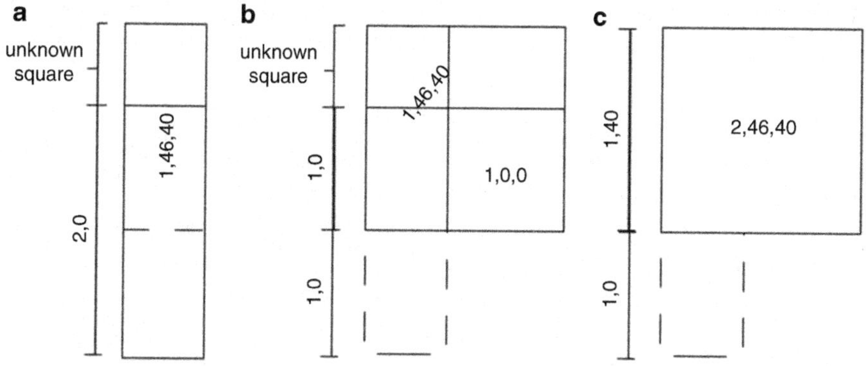

Fig. 4.8 (a–c) The problem about a square solved in Problem 1 of IM52301

to be 1,40) and 20 (the excess of the upper width over the lower width), obtaining 2,0. He now proceeds to solve the problem about a square.

Then, in Fig. 4.8b, the scribe halves the rectangular area of width 2,0 (line 8) and rearranges one of the halves horizontally in relation to the unknown square, a movement that produces an L-shaped region and a square of area 1,0,0 fitting into it. Thus, a new square of area 2,46,40 is formed, equivalent to the sum of the given 1,46,40 and the area 1,0,0 (lines 8 and 9). In the next step, in Fig. 4.8c, the scribe obtains the side of the square of area 2,46,40, namely, 1,40 (line 9). Adding 1,0 to this number, he finally arrives at one of the sides of the original rectangle: 1,40 + 1,0 = 2,40 (lines 10 and 11).

The side 2,40 of the original rectangle, as already said, is thought of as the accumulation of the lengths of the quadrilateral. By subtracting the upper length 1,40 from this number, the scribe gets the missing lower length 1,0 (line 12).

By what at first sight seems to be a coincidence, the lower length equals the accumulation of the widths. This can be justified, however, by halving the accumulation of the lengths, which results in 1,20 and by taking its *igi,* which is 0;0,45. Raising 0;0,45 to the area 40,0, one obtains 30, that is to say, half the accumulation of the widths. Thus, the accumulation of the widths itself is 1,0. While copying the text, the scribe left out this explanation, but we cannot assert whether he did so on purpose or not.

The value 1,0 is then divided by 2 (line 13), producing the average width 30. It is thus only necessary to add and to subtract to and from it half of the difference between the widths: 30 plus 10 gives the upper width 40 (line 14) and 30 minus 10 gives the lower width 20 (line 15). The problem is finally solved.

Problem 2

Let us start with a brief recapitulation of the given data and the question of the problem. In a trapezium, the length (thought of as the distance between the bases) equals two thirds of the accumulation of its upper and lower widths (thought of as the

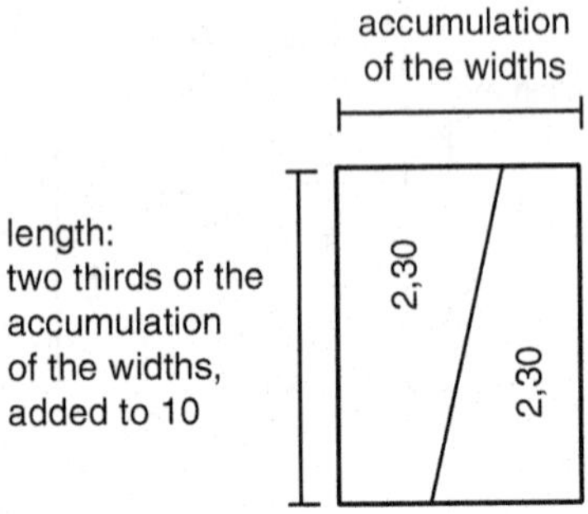

Fig. 4.9 Two copies of the trapezium

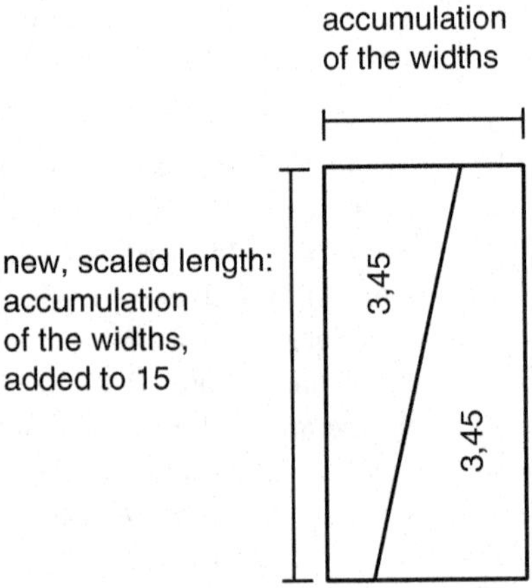

Fig. 4.10 The trapeziums after the scaling along the length

bases) plus 10. Besides, the upper width is bigger than the lower width by 5. The problem also tells that the area of the figure is 2,30. The text then asks for the value of the length.

Thus, the problem may be viewed as dealing with two unknown numbers, (a) the (upper) length and (b) the accumulation of the widths. Their product is twice the area 2,30 of the quadrilateral. Besides, the length equals two thirds of the accumulation of the widths, to which 10 is added. This is summarised in Fig. 4.9.

In order to "cancel" the two thirds, the scribe makes a scaling in the direction of the lengths. This scaling has ratio equal to the *igi* of 0;40 (the reciprocal of two thirds). Thus, after the scaling along the length, we have trapeziums in which the accumulation of the widths is the same; the new length is this accumulation to which 15 is added and the area is 3,45, as in Fig. 4.10.

The length, or upper length, is expressed in terms of the accumulation of the widths.

Now we have two unknown numbers, namely, the accumulation of the widths and the new length (that is to say, the addition of the accumulation of the widths with 15). We know both the product 7,30 (twice the area 3,45) and the difference 15 of these two numbers. That is what enables the scribe to set up a problem about a

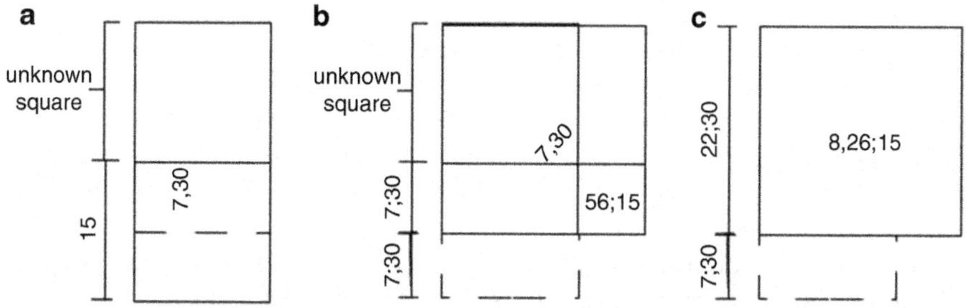

Fig. 4.11 (**a–c**) The problem about a square solved in Problem 2 of IM52301

square corresponding to Fig. 4.11a, where to an unknown square a rectangle of width 15 is added, producing a rectangle of area 7,30 (twice 3,45).

It is solved as follows. In Fig. 4.11b, the scribe halves the rectangular area of width 15 and rearranges one of the halves horizontally in relation to the unknown square, a movement that produces an L-shaped region and a square of area 56;15 fitting into it. Besides, a new square of area 8,26;15 is formed, equivalent to the sum of the given 7,30 and the area 56;15. In the next step, in Fig. 4.11c, the scribe obtains the side of the square of area 8,26;15, namely, 22;30. All this is accomplished in lines R5 to R9.

From 22;30, the scribe removes 7;30 (lines R9 and R10), obtaining the width 15 of the rectangle of area 7,30 of Fig. 4.11a, a number that is therefore the accumulation of the widths of the original trapezium in the problem. After halving the difference 5 between the widths (lines R11 and R12), it is only necessary to add and to subtract this half-difference 2;30 to and from 7;30 (half the width 15). Thus, in lines R12 and R13, the scribe calculates 2;30 plus 7;30, which equals 10, the upper width. In lines R13 and R14, the computation 7;30 minus 2;30 equals 5, the lower width. Thus, the accumulation of the widths is 15 (line R15), its two thirds are 10 (line R16), which added to 10 produces 20, the initially unknown upper length (line R17). Finally, a checking is made: half the sum of the widths is 7;30, a value that raised to 20 gives 2,30, the area (lines R17 and R18). As the scribe writes, this is the procedure.

As one can see, it is not said explicitly which geometrical shape is dealt with. Although there has been a proposal that it is a triangle divided by a segment (Bruins 1951), the prevalent opinion is that it is a right trapezium. This is in accordance with the area calculation in lines R17 and R18, where one can conclude that the sides referred to as widths are the parallel ones and that the distance between them equals the upper length (Høyrup 2002, 215).

4.2.6 Mathematical Interpretation of the So-Called Problem 3

This has been firstly interpreted in a convincing way by Friberg (2000, 118). The text seems to convey a formulation of the general rule for computing areas of quadrilaterals known as the Surveyor's Formula.

Lines E1, E2 and E3 are split in two segments each, as if that section of the text were written in two columns. In the first segment of line E1, the scribe states the problem, namely that of "an area", that is made of "lengths that are not equal to each other". In order to solve the problem, one must know the "sum of the lengths" (first segment of line E1). The scribe is not clear here, but if the interpretation is right this refers to knowing the sums of the opposed lengths. This "sum" is referred to by a different term for additions (see Chap. 2).

The surveyor's formula multiplies the average width by the average length. Equivalently, it divides the product of the sum of opposite sides by four. This enables us to understand the first segment of line E3 and the second segment of line E1, where the reciprocal of 4 is obtained.

In the second segment of line E2, this is multiplied by the sum of the lengths or the sum of the widths in Friberg's (2000, 118) reading. The scribe seems to be combining the previous calculations. Now that he has the result of the product of the sums of opposite sides divided by 4, he declares that "from its interior, you remove the area"(second segment of line E3 and line E4). It is not clear the meaning of "to remove" here, but it is certainly linked to "calculating" the area. Perhaps the scribe used a form of the verb *mašāḫum*, "to measure, to compute" (as in Friberg's reading), but the verb is not otherwise attested for the Old Babylonian period (CAD M1, 352; AHw II, 623).

List of Coefficients

6 40 of the coefficient (of) a basket.

6 of a storehouse.

Robson (1999, 116–117) reports that three possible interpretations have been given to coefficients like these.

These coefficients would be two among several giving the ratio between the capacity in $sila_3$ of a prismatic deposit and the volume in sar_v (that is to say, $nindan^2 \times cubit$) of a cylindrical deposit. In the specific case of the coefficient 6,40,0, the prism has a square base with sides equal to the diameter of the cylindrical deposit. However, other values present in coefficient tables, such as 6,0,0, do not seem to correspond to any obvious shape for the prism.

Secondly, they would be variations of the standard relation 1 $sar_v = 5,0,0$ $sila_3$.

Finally, these coefficients would be two among several (in the range 6,0,0 to 6,45,0, as attested in a Late Babylonian list) that are commodity specific and whose meaning, however, is not known. Speculatively, they could be related to the granularity (maybe density) of the stored thing (grains, oil, water, …).

3 45 a brickwork.

The coefficient 0;3,45 is the volume of a brick wall, in sar_v, a man is able to build in 1 day. In the Tell Harmal mathematical texts, it is used in IM53961 and IM54011.

4 10 of {a brick/bricks}.

There are two possible interpretations for this coefficient. Firstly, it might be understood as the volume of earth a man is capable of carrying. This volume corresponds to six bricks of Type 2: 6×(15×10×5 fingers)=6×0;0,41,40 sar_v=0;4,10 sar_v (Robson 1999, 87).

Another possible interpretation is that 0;4,10 sar_v is the volume of bricks a man is able to pile during 1 day (Robson 1999, 91).

5 of a circle.

0;5 is the coefficient used to transform the square of the circumference of a circle into its area (Robson 1999, 36).

30 of a triangle.

When computing the area of a triangle, A=0;30 width×height (Robson 1999, 41).

7 30 the pile of barley.

This seems to be the conversion 1 sar_v=7,30,0 $sila_3$ (thus one and a half times larger than the standard 1 sar_v=5,0,0 sila). Its usage is not understood, but a similar value, 8,0,0, is used to calculate the capacity of a pile of grain in TMS 14[9] (Robson 1999, 199–120).

4.2.7 *Final Remarks*

All in all, there is no doubt that this is a complex tablet and a rich source for our understanding of Mesopotamian mathematics. It brings a list of coefficients, two variations of a problem and something that might be an effort towards a general understanding of this type of question. The solutions of the problems are also very rich, bringing elaborate examples of the cut-and-place procedure to solve problems about squares, the use of scalings and a conscious effort to check the results numerically at the end of Problem 2. Finally, Problem 1 even seems to mention a physical computation device, through the metaphorical use of the word "hand", as commented above. This all shows that the scribe who wrote it mastered a great deal of Mesopotamian mathematical techniques. Unfortunately, it is an isolated specimen in the mathematics of Šaduppûm, so that all generalisations we could derive from its analysis must be taken with a grain of salt.

4.3 IM54478

4.3.1 *References, Physical Characteristics and Contents*

The tablet was originally published by Baqir (1951). von Soden (1952) brought a pair of philological corrections. The tablet was found in room 252, during the fourth season of work, in 1949. It measures 7.3×5.5×2.2 cm. Robson (1999, 113;

[9] That is to say, Text 14 of Bruins and Rutten's TMS.

2000, 35–36) presented an analysis of the solution of the problem and a complete transliteration. Proust (2007, 218ff) also gave a complete transliteration (equivalent to that published by Baqir with the amendments by von Soden) and a new mathematical interpretation.

4.3.2 Transliteration and Transcription

Obverse

(1) *šum-ma ki-a-am i-ša-al-ka um-ma šu-*u_2*-ma*
šumma kīam išâlka umma šū-ma
(2) *ma-la uš-ta-am-ḫi-ru* u_2*-ša-*pi_2*-il-ma*
mala uštamḫiru ušappil-ma
(3) *mu-ša-ar* u_3 *zu-*$uz_4$$^{?}$ *mu-ša-ri*
mušar u zūz mušari
(4) *e-*pe_2*-ri a-su-uḫ ki-ia uš-tam*$^{?}$*-ḫi-ir*
eperī assuḫ. kīa uštamḫir.
(5) *ki ma-*$ṣi_2$ u_2*-ša-*pi_2*-il*
kī maṣi ušappil.
(6) *at-ta i-na e-*pe_2*-ši-ka*
atta ina epēšika
(7) [1.30 u_3]$^{?}$ 12 *lu-pu-ut-ma i-gi* 12 *pu-*$ṭu_2$*-ur-ma*
1.30 *u* 12 *luput-ma. igi* 12 *puṭur-ma*
(8) [5 *a-na* 1].30 *e-*pe_2*-ri-ka*
5 *ana* 1.30 *eperika*

Reverse

(R1) *i-ši-ma* 7.30 *ta-mar* 7.30
išī-ma. 7.30 *tammar.* 7.30
(R2) *mi-nam* $ib_2.si_8$ 30 $ib_2.si_8$ 30 *a-na* 1
mīnam $ib_2.si_8$ 30 $ib_2.si_8$ 30 *ana* 1
(R3) *i-ši-ma* 30 *ta-mar* 30 *a-na* 1 *ša-ni-im*
išī-ma 30 *tammar.* 30 *ana* 1 *šanîm*
(R4) *i-ši-ma* 30 *ta-mar* 30 *a-na* 12
išī-ma 30 *tammar.* 30 *ana* 12
(R5) *i-ši-ma* 6 *ta-mar* 30 *mi-it-ḫa-ar-ta-ka*
išī-ma 6 *tammar.* 30 *mitḫartaka.*
(R6) 6 *šu-pu-ul-ka*
6 *šupulka.*

4.3.3 *Philological Commentary*

In line 3, we have *mušar* in the status absolutus, *zūz* in the status constructus and *mušari* in the genitive. Finally, in line 4, *e-pe*$_2$*-ri* is to be understood in the case required by the verb, accusative, so it is read *eperī*, my volume (Table 4.3).

In line 8, von Soden (1952, 50) suggests [5 *ta-mar* 5 *a-na* 1]. However, the spacing of the signs seems not to allow this.

4.3.4 *Translation*

[1]If (someone) asks you thus, (saying) this: [2]as much as I caused it to confront itself, so I excavated, and [3,4]I removed a sar$_v$ and half a sar$_v$ of my volume. How did I cause it to confront itself? [5]How much did I excavate? [6]You, in your doing, [7]record 1;30 and 12. Detach the *igi* of 12, and [8, R1,R2,R3,R4,R5]raise 0;5 to 1;30, your volume. You see 0;7,30. What does 0;7,30 make equal? It makes 0;30 equal. Raise 0;30 to 1, and you see 0;30. Raise 0;30 to a second 1, and you see 0;30. Raise 0;30 to 12, and you see 6. Your confrontation is 0;30. [R6]Your depth is 6.

4.3.5 *Mathematical Commentary*

In this problem, a cubic volume is excavated. We know that a cube is dealt with because the geometrical figure corresponding to the excavation has a square base ("… I caused it to confront itself …" (line 2)), and the side of the base is equal to the excavated depth ("as much as" the side of the base, "so I excavated" the depth (line 2)). We are also said that this cube has volume equal to one and a half sar$_v$ (lines 3 and 4).

As, by definition, 1 sar$_v$ is the volume of a right prism with square base of area 1 nindan^2 and height 1 cubit, the scribe immediately converts the height from cubit to nindan. This is done by calculating the *igi* of 12 (line 7), once 1 nindan is composed of 12 cubits. So 1;30 raised to 0;5 (the *igi* of 12) gives 0;7,30, the volume of

Table 4.3 Divergent editorial readings in IM54478

Line	Sign(s)	Divergence
1	*šum*	Printed *šu* in Baqir (1951), obviously a typo
3	*uz*$_4$$^?$	Read uz$_2$ by Baqir (1951)
4	*ki-ia*	Read ki-*ia,* that is to say, *qaqqarija*, by Baqir (1951)

the cube in nindan^3 (lines 8, R1). In the sequence, the scribes asks for the side of the confrontation that originated the cube of volume 0;7,30 and obtains 0;30 (line R2). Thus, the cube that was excavated has sides 0;30 nindan. It is necessary now to convert this result to the original units. As the units for the length and width of the base are the nindan itself, the ratio of conversion for these is 1. Accordingly, the scribe raises 0;30 to 1, obtaining 0;30, which is one of the dimensions of the base (lines R2 and R3); then he raises 0;30 to a second 1, obtaining again 0;30, the other dimension of the base (lines R3 and R4). In order to calculate the depth in cubits, the original unit, the scribe raises 0;30 to 12, and obtains 6 cubits (lines R4 and R5). Finally, he gives the answers: 0;30 nindan is your confrontation (line R5), that is to say, the side of the square base of the excavation, and 6 cubits is your depth (line R6).

An alternative explanation is given by Proust (2007, 218ff). First of all, 1.30 is to be understood as an abstract number, without either a unit of measure attached to it or a specified order of magnitude. Secondly, the conversion of "a sar_v and half a sar_v of volume" to 1.30 is made by consulting the composite table of surfaces.[10] Then it follows a comparison between the volume 1.30 and 12. Here too, 12 is an abstract number (the abstract number that corresponds to the volume of a reference cube, of length 1 nindan, width 1 nindan and height 1 nindan). The ratio between these volumes is calculated by raising 1.30 to the *igi* of 12, resulting in 7.30 (lines 7 to R1). As a consequence, the scribe is able to ask for what 7.30 makes equal (a cube root, in our terms), obtaining 30, the ratio between the lengths of the cubes of volumes 1.30 and 12 (line R2). In the following lines, the scribe uses this ratio to compute the dimensions of the original cube (lines R2 to R5), by multiplying 30 by the dimensions of the reference cube: 30 raised to 1 gives 30, 30 raised to a second 1 gives 30, and 30 raised to 12 gives 6. The results—30, 30 and 6—are abstract numbers, but he could make a new consultation of the metrological tables of lengths and heights and obtain the measures.

In my opinion this should be nuanced by the introduction of some device controlling the orders of magnitude, an aspect that is crucial in empirical applications of mathematics. Thus, using a metrological table (or doing it by heart), the scribe converts the volume to 1.30, but controls its order of magnitude by remembering its position in the table (1 and a half sar_v). The same is valid for the remaining numbers. 7.30 is somehow controlled as what we refer as 0;7,30 and 30 as 0;30. Admittedly, we cannot really be sure that the scribe really does this or if he only thinks about the orders of magnitudes after arriving at the result, as Proust (2013) suggests. Anyway, an historiographically sensitive interpretation should strive to bring evidence for a mixture of techniques and practices that enabled scribes, in spite of their writing system for numbers, to harness the magnitude of the results of their computations.

[10] The metrological table of surfaces was used for volumes too.

4.4 IM53953

4.4.1 References, Physical Characteristics and Contents

The tablet was originally published by Baqir (1951). von Soden (1952) added some philological comments. None of the authors, however, attained a complete understanding of the text, because mainly of some heavily damaged steps in the text, as I explain in the mathematical commentary. The tablet was found in room 252, during the fourth season of work, in 1949. It measures 7.5 × 5.8 × 2 cm.

4.4.2 Transliteration and Transcription

Obverse

(1) *šum-ma ki-a-am i-ša-al um-ma šu-u*$_2$*-ma*
šumma kīam išâl umma šū-ma
(2) *sa-ta*$^?$*-ku-um ši-ni-ip* uš *e-li-im* uš *ša-ap-lu-um*
sattakum. šinīp šiddim elîm šiddum šaplûm.
(3) *is*$^?$ [x x] sag.ki *ša-ap-li-tim* sag.ki *e-li-tum*
[…] *pūtim šaplītim pūtum elītum.*
(4) a.ša$_3$ 2.5 uš *u*$_3$ sag.ki *mi-nu-um*
eqlum 2.5. *šiddum u pūtum mīnum.*
(5) [*at-ta*] *i-na e-pe*$_2$*-ši-ka* 1 u$_3$ ša.na.bi
[*atta*] *ina epēšika* 1 *u šinipêtim*
(6) *ku-mu-ur ḫi-pe*$_2$*-e-ma* 50 *i-li* 50 *ša i-li-a-ku-um*
kumur. ḫipē-ma 50 *illi.* 50 *ša illiakkum*
(7) [… x] *ki* (*di*?) *ḫi-pe*$_2$*-e-ma* 10 *i-li* 10 *ša i-li-a-ku-um*
[…] *ḫipē-ma* 10 *illi.* 10 *ša illiakkum*
(8) *a-na* 50 *i-ši-ma* 8.20 *i-li* 8.20
ana 50 *išī-ma.* 8.20 *illi.* 8.20
(9) [*ša i*]*-li-ku-um* igi *pu-ṭu*$_2$*-ur-ma*
ša illikum igi puṭur-ma

Reverse

(R1) 7.12 *i-li* 7.12 *ša i-li-ku-um*
7.12 *illi.* 7.12 *ša illikum*
(R2) *a-na* 2.5 a.ša$_3$ *i-ši-ma* 15 *i-li*
ana 2.5 *eqlim išī-ma* 15 *illi.*
(R3) 15 *ša i-li-ku-um mi-na-am* [i]b$_2$.si.e
15 *ša illikum mīnam* ib$_2$.si.e.
(R4) 30 ib$_2$.si.e *na-as*$_2$*-ḫi-ir* [x u]š *e-lu-um*
30 ib$_2$.si.e. *nasḫir.* [… *ši*]*ddum elûm.*
(R5) 20 sag.ki *ša-ap-li-tum* 10$^?$ sag.ki *e-li-tum*
20 *pūtum šaplītum.* 10 *pūtum elītum.*

4.4.3 Philological Commentary

It is also worth mentioning that in this text, as in IM52301, the word for triangle is written syllabically, and it leads to *sattakkum* (Table 4.4).

In lines 8 and 9, the scribe seemed to mean the *igi* of 8.20. It remains to explain why igi and 8.20 are written separately, for this is quite unusual.

von Soden (1952) makes a number of interesting remarks:

- He sees a remaining part of line 6, which he suggests to be *lu-pu-ut*. This would match three possible very small signs written on the edge of the tablet, which I omitted in the transliteration and transcription above.
- Perhaps the beginning of line 7 is sag.ki, although this would not really make sense, for the problem deals with two unknown widths.
- The photograph causes an impression that there was a half line written between lines R4 and R5 on the left side of the tablet's surface; perhaps this half line dealt with the calculation of the lower length.

Finally, I must mention that the beginnings of lines 3 and 7 are not referred to in the vocabulary.

4.4.4 Translation

[1]If (someone) asks thus (saying) this: [2]a triangle. Two thirds of the upper length is the lower length. [3][...] of the lower width is the upper width. [4]The area is 2.5. What are the length and the width? [5,6]You, in your doing, accumulate 1 and two thirds. Halve it, and 50 comes up. 50 that comes up to you, [7][...] Halve it, and 10 comes up. 10 that comes up to you [8,9]raise to 50, and 8.20 comes up. Detach the *igi* of 8.20 that comes up to you, and [R1,R2]7.12 comes up. Raise 7.12 that comes up to you to 2.5, the area, and 15 comes up. [R3]What does 15, that comes up to you, make equal? [R4]It makes 30 equal. Return. The upper length is [...] [R5]The lower width is 20. The upper width is 10.

Table 4.4 Divergent editorial readings in IM53953

Line	Sign(s)	Divergence
2, 3, R5	*ap*	Read ap_2 by Baqir (1951)
5	bi	Read pi_2 by Baqir (1951)
6	50 *i-li* 50 *ša*	Read 50 by Baqir (1951), most probably only a misprint
R3	*i-li-ku-um*	Read *i-li-a-ku-um* by Baqir (1951)
R4	*e-lu-um*	Omitted by Baqir (1951)

4.4.5 Mathematical Commentary

As mentioned above, no clear understanding of the problem has been reached by the field. Because of this, it is not possible to give a coherent understanding of the calculations involved in the text. Especially, it is not possible to assign the relative orders of magnitude of the numbers that are present in the text. However, in order to help the reader unaccustomed with floating point calculations, I assume some relative orders of magnitude, but we should keep in mind that this assumption is an arbitrary one. Furthermore, even if we knew the relative orders of magnitudes, it would remain the possibility of not being able to determine the absolute orders of magnitude of the numbers in this problem, as exemplified in the end of the commentary to tablet IM55357.

The statement of the problem begins by telling us that a triangle is the object of this text, although four measures are mentioned. Specifically, information is given about the lengths, namely, that the lower length equals two thirds of the upper length (line 2). In line 3, there must have been data regarding the widths, for this line seems to relate a part (the missing signs at the beginning of the line) of the lower width to the upper width. However, as the first, perhaps three signs are damaged to the point of not being readable, it is not possible to know what the scribe wrote here. von Soden (1952, 51) suggested that some word for "half" should have been present in the beginning of line 3, for in the solution to the problem it seems to be written that the lower width is 20 and the upper width is 10. The supposition is sound, although von Soden himself agreed that what remains of the signs does not permit the reading of any known Akkadian word with the meaning of half. Finally, the scribe tells us that the area is 2,5 (line 4).

In my opinion, the problem deals with a triangle that is divided by a segment, as in Fig. 4.12. The following analysis refers to the geometrical elements of this figure. The area, according to the present interpretation, is the area of the quadrilateral formed in the left side of the figure.

The scribe begins the solution by accumulating 1 and two thirds, that is to say, 1 and 0;40, numbers that are proportional to the lengths, thus indicating either that

(I) The procedure is that of the false position or
(II) The scribe is able to set up another figure, similar to that of the problem, and to compare both. This second figure would be the result of an application of horizontal and vertical scalings with the same ratio to the triangle sketched above[11]

Next he breaks the result in half (lines 5 and 6), which is a way of computing an average length, or rather, a hypothetical length that is proportional to the actual average length. The result is 0;50 and it appears again in lines 7 and 8 raised by 0;10, the last value being probably related to the widths of the figure. Because 0;10 was obtained as the result of breaking in half something that is not readable any more in the beginning of line 7, it is possibly an average width, which would suggest

[11] See also the commentary to IM54478, where the strategy of positing a second figure, similar to that of the problem, might have been in action too.

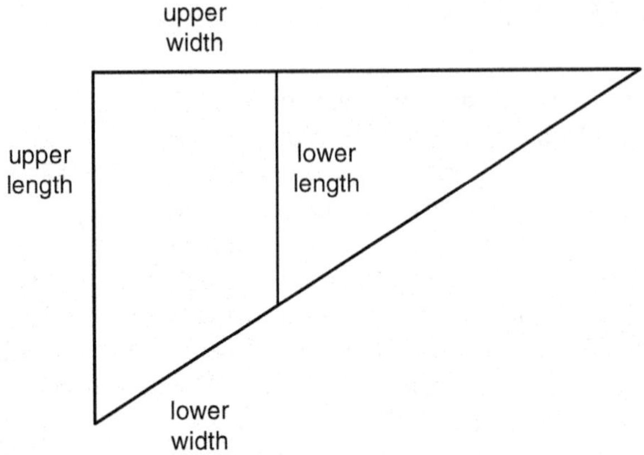

Fig. 4.12 A possible configuration for IM53953

that the scribe is using the well-known approximate formula for the area of a quadrilateral (as we saw in IM52301), equaling the area to the product of the averages of opposite sides.

Now we must notice that it would make sense as well to assume, as I do in Fig. 4.12, that the scribe was dealing with a trapezium of height 0;10, where the height is to be understood, in the scribe's terms, as the upper width. Thus, assuming also that the upper width is half the lower width (which therefore must equal 0;20), then in line 7 the scribe would break the lower width in half, obtaining the upper width, necessary to compute the area of such a trapezium. One problem with this interpretation is that it does not account for the value of the lower width, namely 0;20: how did the scribe set it? Is it thought of as a proportional measure, like the numbers 1 and 0;40 that were taken for the lengths?

Anyway, the result of the last raising is 0;8,20. The scribe proceeds at this point to the computation of the ratio of the given area 2,5 to the obtained area 0;8,20: the *igi* of 0;8,20 is 7;12 (line R1), and 7;12 raised to 2,5 equals 15,0 (line R2). As this is a ratio between areas, its square root is a ratio between linear measures, which seems to be coherent with hypothesis (II) above. This ratio is obtained in lines R3 and R4: 15,0 causes 30 to be equal. Thus, still under hypothesis (II), 30 is the similarity ratio between the figure of the problem and the second figure set up by the scribe (or the ratio of the horizontal and vertical scalings).[12]

Unfortunately, the second half of line R4 is damaged. We can only read that something about the upper length is said here, but it is not possible to know what. This, with the already mentioned difficulty of reading in line 3, makes it hard, if not impossible to know exactly what is going on in this problem.

Anyway, we could then expect that the ratio 30 would be used to calculate the lengths and the widths of the quadrilateral. If this were the case, the upper length of the sought figure would be 1 raised to 30, equaling 30. The lower length would be 0;40 raised to 30, giving 20. The width 0;10 in the scaled figure would correspond to 0;10 raised to 30, which is equal to 5.

[12] See again IM54478, where a similar computation of a linear ratio from the ratio of volumes might have been present.

In the last lines of the text (lines R4 and R5), however, we read that

- The upper length is [x] in the transliteration, which would make x equal to 30, but to be confirmed (if so) only by collation.
- The lower width is 20, but in reality it was the lower length that we expected to be 20.
- The upper width is 10, which cannot be made compatible with the number 5 we expected for it; in fact, we expected the lower width to be 10 and not the upper width.

Evidently, all this requires a considerable amount of supposition. In favour of it, there are the facts that [x], in line R4, is not readable, and 10 (line R5), the value of the upper width, is a dubious reading. But this supposition also requires that we change "lower width" (line R5) to "lower length" and "upper width" to "lower width", which is a heavy, if not unreasonable demand.

These handicaps apart, it should finally be noticed that this tablet has three interesting points of contact with IM52301. Firstly, the problem in IM53953 and Problems 1 and 2 in IM52301 deal with a quadrilateral, although in IM53953, the quadrilateral is set inside a triangle. Secondly, in the three problems, the widths are such that one is half the other. Finally, as already mentioned in the previous philological commentary, both tablets also write *sattakkum* and not *santakkum*.

4.5 IM54538

4.5.1 References, Physical Characteristics and Contents

The tablet was originally published by Baqir (1951). von Soden (1952) suggested some improvements on the reading, as well as a clearer explanation for its contents. Additional commentaries on the mathematics of this text are found in Robson (1999) and Friberg (2001). However, a precise understanding of the problem is still lacking. The tablet was found in room 252, during the fourth season of work, in 1949. It measures 8.6×6×2.5 cm.

4.5.2 Transliteration and Transcription

Obverse

(1) *šum-ma ki-a-am i-ša-al-ka* [*um-ma*] *šu-u*$_2$*-ma*
šumma kīam išâlka umma šū-ma
(2) *a-ša-al ši-du-um e-še-re-*[*et m*]*u-*[*ša*]*-ar*!
ašal šiddum. ešeret mušar

(3) *li-bi-tu-um i-na* ki.su$_7$-*im ša-*[*ak-na*]*-at*
libittum ina maškanim šaknat.
(4) *ki ma-ṣi*$_2$ *ṣa*$_2$*-ba-am u*$_2$*-ma-ka-li-a-am*
kī maṣi ṣābam ūmakaliam
(5) *lu-uš-ku-un-ma li-ig*!*-mu-ra-am*
luškun-ma ligmuram.
(6) *at-ta i-na e-pe*$_2$*-ši-*[*k*]*a* 1.30 *i-gi-gu-ub*$^{?}$*-bi*
atta ina epēšika 1.30 *igigubbî*
(7) *šu-ku-un-ma i-gi* 1.30 *i-gi-gu-bi-ka*
šukun-ma igi 1.30 *igigubbîka*
(8) *pu-ṭu*$_2$*-ur* 40 *ta-mar* [40 *a*]*-na a-ša-al*
puṭur. 40 *tammar.* 40 *ana ašal*

Edge

(E9) *ši-di-im i-ši-ma* 6.40 *ta-mar*
šidim išī-ma 6.40 *tammar.*
(E10) 6.40 *ša ta-mu-ru a-na* 54
6.40 *ša tammuru ana* 54

Reverse

(R1) *e-še-re-et mu-ša-ri li-bi-ti-ka*
ešeret mušari libittika
(R2) *i-ši-ma* 6 *ta-mar* 6 *a-wi-lu-ka*
išī-ma 6 *tammar.* 6 *awīlūka*
(R3) *u*$_2$*-ma-ka-lu-tu-un*
ūmakalūtun
(R4) *ša i-ga*!*-ma-ru-ni-iš-ši*
ša igammarūnišši
(R5) *i-na u*$_2$*-ma-ka-al*
ina ūmakkal.

4.5.3 Philological Commentary

In lines 5 and 7, it is possible to assimilate the final n to m: *luškun-ma* and *šukun-ma* (Table 4.5).

In line R1, we should expect the status absolutus *mušar*. As a matter of fact, Baqir (1951) transliterated this word as *mu-ša-ri* both in lines 2 and R1. von Soden (1952) explicitly suggested a correction on the scribe's writing for line 2. For the sake of consistency, this correction applies to line R2 too. Anyway, it is worth noticing the scribe's usage, *mu-ša-ri*, deviating from the expected status absolutus (see also GAG §62d).

Table 4.5 Divergent editorial readings in IM54538

Line	Sign(s)	Divergences
2	*[m]u-[ša]-ar!*	Read *[mu-ša-r]i* by Baqir (1951)
3	ki.su_7	Read *ki-di*(?) by Baqir (1951)
3	*ša-[ak-na]-at*	Read *ša x x*(?)*-at*(?) by Baqir (1951)
4	u_2*-ma-ka-li-a-am*	Read u_2*-ma ka-li-a*(?)*-am* by Baqir (1951)
5	*li-ig!-mu-ra-am*	Read *li-[iš ?] mu-ra-am* by Baqir (1951)
R3	u_2*-ma-ka-lu-tu-un*	Read u_2*-ma ka-lu-tu-un*(?) by Baqir (1951)
R4	*i-ga!-ma-ru-ni-iš-ši*	Read *i-ša-ma-ru ni-iš-ši* or *i-ša-ma-ru-ni iš-ši* by Baqir (1951)

4.5.4 *Translation*

[1]If (someone) asks you thus, (saying) this: [2]one rope is the distance. Ten sar [3]of bricks are placed on the threshing floor. [4]How many workers for 1 day [5]should I place so that they finish (the task)? [6,7,8,E9]You, in your doing, place 1,30,0, my coefficient, and detach the *igi* of 1,30,0, your coefficient. You see 0;0,0,40. Raise 0;0,0,40 to one rope of length and you see 0;0,6,40. [E10,R1,R2]Raise 0;0,6,40 that you see to 54,0, ten sar of your bricks, and you see 6. Your workers for 1 day are 6, [R4]who finish it in [R5]1 day.

4.5.5 *Mathematical Commentary*

The problem establishes that a definite quantity of bricks must be transported from a threshing floor area that is located at a certain distance (line 3): specifically, ten sar of bricks to be transported over a distance of one rope. So, let us start by explaining these data. "Ten sar of bricks" might be understood to begin with in two different senses. The expression might refer to a volume, that is to say, 10 $\text{sar}_v = 10$ nindan$^2 \times$cubit of bricks. Alternatively, it might refer to a standard amount of bricks, the so-called brick sar, indicated by sar_b and corresponding to an amount of 12,0 bricks. In the latter case, 10 $\text{sar}_b = 10 \times 12{,}0 = 2{,}0{,}0$ bricks (that is to say, 7200 bricks). As for the given distance, it is known that a rope corresponds to 10 nindan or 2,0 cubits (that is to say, 120 cubits).

As the problem deals with the carrying of bricks, we would expect that the solution used one of the coefficients specific for carriage, Akkadian *nazbalum*. The carriage coefficient is the number of bricks that one man is able to transport over a distance of 1 nindan in 1 day (or over a distance of 30 nindan in 1 month of 30 days, which is the same). Bricks with dimensions $30 \times 5 \times 15$ fingers, the so-called type 8a (Robson 1999, 71), have the *nazbalum* coefficient equal to 1,30,0 (that is to say, 5400 bricks). As this number appears in the problem, we can temporarily assume that it is this kind of bricks that is dealt with in the problem.

As the given distance is 10 nindan (1 rope), the scribe multiplies 10 nindan by the *igi* of 1,30,0, obtaining 0;0,6,40 (which is the "fractionary number" of workers necessary for the task of transporting only one brick over a distance of 10 nindan). We would expect now that the scribe multiplied the last value by 10 sar_b, namely, 2,0,0. Instead, he uses a value described as "54,0, ten sar of your bricks", obtaining the result of 6 necessary men for the task (0;0,6,40 raised to 54,0). Let us notice that, if the scribe used the value 2,0,0, the result would be 13;20 men, that is to say, 13 workers and one third of a worker!

Robson (1999, 84) pointed out that, if the text is read with the meaning of 54 $sar_b = 10\ sar_v$, then 1 $sar_v = 5;24\ sar_b$, implying that the bricks are of a different type, having dimensions $20 \times 10 \times 5$ fingers and "a *nazbalum* of 3,22,30, not 1,30,0, as the text seems to suggest". Although this does not solve the problem of interpretation that this tablet brings, it is a possibility to keep in mind.

A third way of reading "Ten sar of bricks" gives us a better understanding of the problem, as in von Soden (1952) and Friberg (2001). In the present problem, the scribe is referring to a surface-brick-sar, that is to say, the number of bricks that cover a surface of 1 sar (1 nindan $\times$ 1 nindan).

In fact, square bricks of sides 20 fingers (2/3 cubit) are such that 324 pieces cover 1 sar:

$$324 \times 2/3\text{cubit} \times 2/3\text{cubit} = 18 \times 2/3\text{cubit} \times 18 \times 2/3\ \text{cubit} =$$
$$12\ \text{cubit} \times 12\ \text{cubit} = 1\ \text{nindan} \times 1\ \text{nindan}$$

In this way, 10 sar of bricks make 3240 bricks, an amount that can be written in sexagesimal notation as 54,0, which explains the expression "54,0, ten sar" that the scribe used.

These bricks can be the so-called type 8 (Robson 1999, 71) and S6 (Friberg 2001, 79), with thickness five fingers, or type 9 (Robson 1999, 71) and S6v (Friberg 2001, 80), with thickness 6 fingers.

However, a problem remains: neither of these brick types has a *nazbalum* of 1,30,0. Type 8 corresponds to 1,41,15 and type 9 to 1,24,22;30. Friberg (2001, 99) notices that "the discrepancy is not very large" and he suggests that 1,30,0 "may be a deliberately round number close" to the actual coefficient.

4.6 IM53961

4.6.1 References, Physical Characteristics and Contents

The tablet was originally published by Baqir (1951). von Soden (1952) introduced a few corrections in the transliteration. Robson (1999) also gave a complete transliteration, a translation and a commentary. The tablet was found in room 252, during the fourth season of work, in 1949. It measures $6.5 \times 5 \times 2$ cm.

4.6.2 Transliteration and Transcription

Obverse

(1) *šum-ma ki-a-am i-ša-al-ka um-ma šu-u*$_2$*-[ma]*
šumma kīam išâlka umma šū-ma
(2) *pi*$_2$*-ti-iq-tum ši-ta am-ma-tim*
pitiqtum. šittā ammātim
(3) *ru-up-šu-um am-ma-at me-li-um*
rupšum. ammat mēlium.
(4) *iš-ka-ar iš-te-en a-wi-li-im*
iškar ištēn awīlim
(5) *mi-nu-um at-ta i-na e-pe*$_2$*-ši-ka*
mīnum. atta ina epēšika
(6) [10] *a-na* 1 *me-li-ka i-ši-i-ma*
10 ana 1 *mēlika išī-ma*
(7) 10 *i-li* 10 *ša i-li-a-ku-um*
10 *illi.* 10 *ša illiakkum*

Reverse

(R1) *i-gi* 10 *pu-ṭu*$_2$*-ur-ma* 6 *i-l[i]*
igi 10 *puṭur-ma* 6 *illi.*
(R2) [6] *ša i-li-kum a-na* 3.45 *i-gi-gu-bi-ka*
6 *ša illikum ana* 3.45 *igigubbîka*
(R3) [*i*]*-ši-i-ma* 22.30 *i-li* 22.30 *ša i-li-kum*
išī-ma 22.30 *illi.* 22.30 *ša illikum*
(R4) *iš-ka-ar iš-te-en*
iškar ištēn
(R5) *a-wi-li-im i-li*
awīlim illi.

4.6.3 Philological Commentary

Table 4.6 lists the divergent readings.

For the expression *šittā ammātim*, see Sect. 2.4.

4.6.4 Translation

[1]If (someone) asks you thus, (saying) this: [2,3]a brickwork. The width is two cubits. The height is one cubit. [4,5]What is the work of one man? You, in your doing, [6]raise 0;10 to 1, your height, and [7]0;10 comes up. 0;10 that comes up to you: [R1]detach

Table 4.6 Divergent editorial readings in IM53961

Line	Sign(s)	Divergence
2	*ši-ta*	Read *ši-ta-x* by Baqir (1951)
2	*am-ma-tim*	Read *qa-ta-tim* by von Soden (1952)
6	[10] *a-na* 1 *me-li-ka i-ši-i-ma*	Read [2 ?] *a-na* 1 *me-li-ka i-ši-i-ma* by Baqir (1951)
R3	[*i*]-*ši-i-ma*	Read [*i*]-*ši-ma* by Baqir (1951)

the *igi* of 0;10, and 6 comes up. [R2,R3]Raise 6 that comes up to you to 0;3,45, your coefficient, and 0;22,30 comes up. 0;22,30 that comes up: [R4]the work of one [R5]man comes up.

4.6.5 *Mathematical Commentary*

This problem deals with the construction of a brick wall, the cross section of which is 2 cubits wide and 1 cubit high. It asks for the length of wall a man is able to build in 1 day.

In the solution of the problem, the scribe resorts to the coefficient 0;3,45 (line R2), which is known to give the daily volume of a brick wall that one man is capable of building. As this volume is given in sar_v, the scribe has to use the value 0;10 nindan for the width, instead of the equivalent 2 cubits of the given data. It is also possible that the scribe consulted a metrological table, thus converting 2 cubits to the abstract number 10, which we can make correspond to 0;10; if this is the case, we can furthermore assume that the scribe converted the height of 1 cubit to the abstract number 1, by using another metrological table, the one of heights.

The scribe starts by calculating the area of the cross section, which in the present case seems to be a rectangle. The width of 0;10 nindan (or the corresponding abstract number 10) is raised to the height of 1 cubit (or the corresponding abstract number 1). This gives 0;10 as the area (lines 6 and 7), but as if in mixed units, nindan × cubit. The portion of wall a man builds in 1 day has volume equal to the coefficient 0;3,45, so the required length is calculated in two steps. The scribe detaches the *igi* of 0;10 (line R1), obtaining 6 and then raises 6 to 0;3,45 (line R2 and R3), which is structurally equivalent to dividing the daily volume by the area of the cross section. The result 0;22,30 nindan is the length of wall that one man builds during 1 day, called here the work of one man (lines R3, R4 and R5). Interestingly, this last value is not converted back to cubits: should it be understood that the scribe deliberately left it in abstract numbers? Or else in nindan? This seems to be a piece of evidence that in some contexts scribes were not interested in making the distinction.

4.7 IM53957

4.7.1 References, Physical Characteristics and Contents

This tablet was firstly published by Baqir (1951). von Soden (1952) proposed alternative readings to a few passages, leading to a different mathematical interpretation. Bruins (1953a) presented a series of arguments supporting Baqir's point of view. **Interpretation 1**, in what follows, is based on the standpoint maintained by Baqir and Bruins. **Interpretation 2** agrees in its essential points with von Soden. The tablet was found in room 252, during the fourth season of work, in 1949. It measures 8×6×2.5 cm.

4.7.2 Transliteration and Transcription

Interpretation 1

Obverse

(1) *šum-ma* [*ki*]-*a*-[*am i-ša-al-ka*$^?$ *um-ma šu-u*$_2$*-ma*]
šumma kīam išâlka umma šū-ma
(2) *a-na ši-ni-ip ši-ni-pi-ia* me sila$_3$ še
ana šinip šinipija meat qa še'am
(3) *u*$_3$ *ši-ni-pi*$_2$ *u*$_2$*-ṣi*$_2$*-im-ma*
u šinipī ūṣim-ma
(4) [*re-še*$_{20}$]-*um i-ta-ak-ma-ar*
rēšum ittakmar
(5) *re-še*$_{20}$*-e-ia mi ki ma-ṣi*$_2$
rēš še'ja mi kī maṣi.
(6) *at-ta i-na e-pe*$_2$*-ši-ka*
atta ina epēšika
(7) ša.na.bi *u*$_3$ ša.na.[bi]
šinipêtim u šinipêtim

Reverse

(R1) *šu-ta-ki-il-ma* 26.40 *ta-mar*
šutākil-ma 26.40 *tammar.*
(R2) 26.40 *i-na* [1] *ta-ba-al-ma* 33.20
26.40 *ina* 1 *tabal-ma* 33.20
(R3) *ši-ta-tum i-gi* 33.20 *pu-ṭ*[*u*$_2$*-ur-ma*]
šittātum. Igi 33.20 *puṭur-ma*
(R4) [1].48 *ta-mar* 1.48 *a*-[*na* 1.40]
1.48 *tammar.* 1.48 *ana* 1.40
(R5) [*i*]-*ši-ma* 3 *ta-mar* 3 *re-še*$_{20}$*-e-im*
išī-ma 3 *tammar.* 3 *rēš še'im.*

Interpretation 2—only the lines where it differs from Interpretation 1

Obverse

(4) [1 gur]-*um i-ta-ag-ma-ar*
1 *kurrum ittagmar*
(5) *tal-<li> še*$_{20}$*-e-ia mi ki ma-ṣi*$_{2}$
talli šeʾja mi ki maṣi.

Reverse

(R5) [*i*]-*ši-ma* 3 *ta-mar* 3 dal *še*$_{20}$*-e-im*
išī-ma 3 *tammar.* 3 *parsikātum talli šeʾim.*

4.7.3 Philological Commentary

Line 4. The verbal form in this line is Ntn preterite, third singular masculine of *kamārum*, without the iterative–repetitive meaning. It is just a passive form.
Line 5. *mi* added to a word indicates direct speech. This is indeed the case here, as the task is presented after the typical direct speech formula in lines 1 and 2. One should notice, however, that *mi* is not a common occurrence in mathematical texts.
Line R2. *tabal-ma*, that is to say, the Gt imperative of *wabālum.*

As regards Interpretation 2, it is possible to add the following:

Line 4. The verbal form in this line is Ntn preterite, third singular masculine of *gamārum*, without the iterative-repetitive meaning. It is just a passive form.
Line R5. According to von Soden's reading, the last number in this line should be considered to express 3 barig (Akkadian *parsiktum*). However, we must bear in mind that, in the Old Babylonian practice, this is frequently indicated by writing the numeral 3 in a specific way, something that the scribe did not do, as von Soden also noticed. Thus, the interpretation relies heavily on an understanding of the problem as a whole. A second remark to be made regarding this line is the following. When we have "number + unit of measurement + thing measured", each term is usually rendered, respectively, in the status absolutus, the status absolutus and the status rectus (with case from context). When we have "number + thing counted", the terms assume, respectively, the status absolutus and the status rectus (with case from context). I take 3 (barig) to be an example of the last case, so *parsiktum* should be written in the status rectus.

Table 4.7 Divergent editorial readings in IM53957

Line	Sign(s)	Divergences in relation to Interpretation 1
4	[*re-še*$_{20}$]-*um*	Read ... -*um* by Baqir (1951)
5	*mi*	Read *x* by Baqir (1951)
R3	*ši-ta-tum*	Read igi-*ta-tum* by Baqir (1951)

4.7.4 *Translation*

Interpretation 1

[1]If (someone) asks you thus, (saying) this: "[2]to two thirds of my two thirds, a hundred sila$_3$ of barley [3]and my two thirds I added, and so [4]the original quantity is accumulated. [5]The original quantity of my barley", I say, "how much?" [6]You, in your doing, [7]two thirds and two thirds [R1]cause them to combine, and 0;26,40 you see. [R2]0;26,40 from 1 you carry off, and 0;33,20 is [R3]the remainder. The *igi* of 0;33,20 you detach, and [R4]1;48 you see. 1;48 to 1,40 [R5]you raise, and 3,0 you see. 3,0 is the original quantity of barley.

Interpretation 2

[1]If (someone) asks you thus, (saying) this: "[2]to two thirds of my two thirds, a hundred sila$_3$ of barley [3]and my two thirds I added and so [4]1 gur is complete. [5]The *tallum-vessel* of my barley", I say, "how much (is it)?" [6]You, in your doing, [7]two thirds and two thirds [R1]cause them to combine, and 0;26,40 you see. [R2]0;26,40 from 1 you carry off, and 0;33,20 is [R3]the remainder. The *igi* of 0;33,20 you detach, and [R4]1;48 you see. 1;48 to 1,40 [R5]you raise, and 3,0 you see. 3 *parsiktu* is the *tallum*-vessel of barley.

4.7.5 *Mathematical Commentary*

Interpretation 1

Although proposed by Baqir (1951) and Bruins (1953a), both pointed out that this interpretation does not account for the expression *u*$_3$ *ši-ni-pi*$_2$ in line 3, the underlined "and my two thirds" in the translation.

The problem asks for the value of an unknown quantity of barley, and its statement gives us the information that the sum of two thirds of two thirds of this quantity with 100 sila$_3$ (=1,40 sila$_3$) of barley is equal to the original quantity. The scribe raises 0;40 to 0;40, in order to obtain the fraction of the original quantity that was added to 100 sila$_3$. This is done in lines 7 and R1, and the result is 0;26,40. Next the scribe subtracts this value from 1 (lines R2 and R3), because the remainder 0;33,20 is the fraction of the original quantity that corresponds to 100 sila$_3$. Finally, the scribe detaches the *igi* of 0;33,20 (line R3), which is equal to 1;48 (line R4), and raises it to 1,40, that is to say, to 100 sila$_3$ (lines R4 and R5). The result, 3,0 sila$_3$, is the original amount of barley (line R5).

The procedure is quite direct, and it is structurally equivalent to solving a first-degree equation.

Interpretation 2

According to von Soden (1952), the statement of the problem informs us that the sum of two thirds of two thirds of an unknown quantity of barley with 100 $sila_3$ of barley and two thirds of the unknown quantity is equal to 1 gur of barley, that is to say, 300 $sila_3$. This unknown quantity of grain is supposed to be the capacity of a specific container (the *tallum*, a certain kind of vessel), so that the problem requires one to find out how much the *tallum* contains.

In order to make the numbers present on the tablet agree with the solution of the problem in the way it is stated, this interpretation assumes that the scribe is able to carry out a transformation on the data, which however is not explicitly written on the tablet. Firstly, he would have noticed that the sum of two thirds of two thirds of the *tallum*-vessel with two thirds of the vessel must be equal to 200 $sila_3$. In other words, the grey portions in the two schematic vessels of Fig. 4.13a, b add to 200 $sila_3$ (the total 300 $sila_3$ from which 100 $sila_3$ are removed).

Consequently, the scribe would be able to observe that the average of the grey portions is equal to 100 $sila_3$ (because they add to 200 $sila_3$). At the same time, this average is equal to the empty part in the first figure. Therefore, it would be possible to state that two thirds of two thirds of the vessel (as in Fig. 4.13a) added to 100 $sila_3$ (its empty part) is equal to the whole vessel, and in this way we are led to exactly the same situation as that of Interpretation 1. Of course, all this reasoning is quite involved and is not explicitly brought out in the tablet, so that one is not able to assert that this is really what the scribe had in mind. On the other hand, the reasoning does not resort to symbolic manipulation, which is a point in favour of it.

One interesting difference that this interpretation has in relation to the previous one is that it leads to an answer equal to 3 and not 3,0. This is in accordance with the fact that 1 *parsiktu* equals 1,0 $sila_3$. Thus, the *tallu*-vessel of my barley contains 3,0 (=180) $sila_3$, that is to say, 3 *parsiktu* (line R5). Whether we should read 3,0, as in Interpretation 1, or 3, as in interpretation 2, the cuneiform writing of numbers cannot say, and this is in fact immaterial, for this difference is simply an artefact

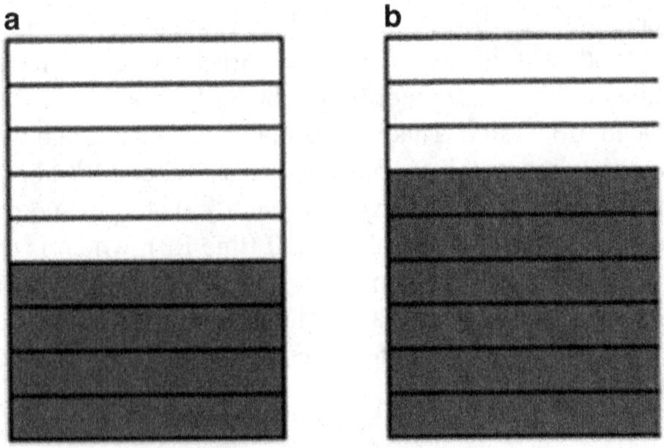

Fig. 4.13 Geometrical representations of the vessel and its fractions. (**a**) Two thirds of two thirds of the vessel. (**b**) Two thirds of the vessel

produced by our use of digit separators: from the point of view of scribes, both 180 sila$_3$ and 3 *parsiktum* correspond to the abstract number 3.

We should also take into account a suggestion by Høyrup (2002, 321), in order to understand the possible aims of this tablet. According to him, the problem might be a "mock-reckoning, a challenge meant to impress and to make fools of the non-initiate". This would explain why there is no clear mathematical reasoning to follow: the scribe would have simply started from the solution. The existence of the strongly parallel problem 37 in the Rhind Mathematical Papyrus would point out to the possibility that both the Egyptian and the Ešnunnan scribes were influenced by similar, if not the same, traditions, producing, however, different kinds of mathematics, a mathematical deductive explanation in the Egyptian text and the permanence of the riddle style in the cuneiform tablet.

4.8 IM54010

4.8.1 References, Physical Characteristics and Contents

Baqir (1951) contains the first publication of IM54010. The tablet is however, very badly preserved and so of difficult, maybe improbable reading. Because of this, Baqir offered neither a translation nor a commentary. von Soden (1952) presented suggestions for improving partially the reading of lines 2–5 (incorporated below), but these were not enough to reach a comprehension of the mathematical meaning of this text. The tablet was found in room 252, during the fourth season of work, in 1949. Its dimensions were not given by Baqir (1951).

4.8.2 Transliteration and Transcription

Obverse

(1) *šum-ma ki-a-*[*am i-ša-al-k*]*a um-ma šu-u*$_2$*-ma*
šumma kīam išâlka umma šū-ma
(2) *er-be*$_2$*-e i-na qa-na ši-ni* kuš$_3$*-ia mi*$^?$ [xx]
erbê ina qanâ [...] *ammatīja* [...]
(3) 2 *qa-a*$^?$ *e-ṣi*$_2$*-id i-na bu-ur ki*$^?$ *ma*$^?$*-ṣi*$_2$$^?$
2 *qâ ēṣid. ina bur kī maṣi*
(4) *e-ṣi*$_2$*-id at-ta i-na e-pe*$_2$*-ši-ka*
ēṣid. atta ina epēšika
(5) *i-gi* {*ir-bi*/*er-be*$_2$} *ši-ni-i-ša pu-ṭu*$_2$*-ur-ma* 15 *ta-mar*
igi {*irbi*/*erbê*} *šinīša puṭur-ma* 15 *tammar.*
(6) 15 *a-na ba-a qa-ni-ka i-ši-ma* 7.30 *ta-mar*

15 *ana bâ qanika išī-ma* 7.30 *tammar.*
(7) [*i*]-*gi* 7.30 *pu-ṭu*$_2$*-ur-ma* 8 *ta-mar*
igi 7.30 *puṭur-ma* 8 *tammar.*
(8) *na-as*$_2$*-ḫi-ir* […] 2$^?$ *qa-ni-ka*
nasḫir […] 2$^?$ *qanika*

Edge

(E9) [x] *ni* [. . ib$_2$].si *mi-na-am* 10$^?$ (40$^?$)
[…] ib$_2$.si *mīnam* 10 (40$^?$)
(E10) *ta-mar* […] *ša*$^?$*-ni-im* [*i*]-*ši-ma*
tammar […] *šanîm išī-ma*

Reverse

(R1) 16 *ta-mar* 16 *a-na* 1 *i-ši-ma*
16 *tammar.* 16 *ana* 1 *išī-ma*
(R2) 16 *a-na bu-ur* a.ša$_3$*-ka i-ši-ma*
16 *ana bur eqelka išī-ma*
(R3) 8 *ta-mar* 8 [x]-*ka*
8 *tammar.* 8 […]
(R4) 8 [xx] *šu ki ma*$^?$*-ṣi*$_2$ *ti*$^?$ 3$^?$
8 […] *kī maṣi* […] 3$^?$
(R5) *u*$_3$ *ši* [x] *ti* [x] *bu-šu*
u […]

4.8.3 Philological and Mathematical Commentaries

The question posed by the problem seems to be *ina bur kī maṣi ēṣid*: "in a *burum*, how much did I harvest?" (lines 3 and 4). Thus, lines 2 and 3 must have contained the given data, but it is not possible to read anything here except a fragmented sentence. The initial word is probably the number four, *er-be*$_2$*-e*, after which it follows *ina* (in, from) *qanâ* (field? reed?), *ammatīja* (of my cubit), 2 *qâ ēṣid* (I harvested 2 sila$_3$). von Soden (1952, 52) discarded the reading of *ši-ni* as the number two. On the other hand, the writing *ši-in-ni* in Problem 7 of Haddad 104 (al-Rawi and Roaf 1984, 202–203), meaning "two", may point to the contrary. However, this alone does not seem to lead to a clearer understanding of the text.

In line 5, the scribe apparently detaches the *igi* of 4, which agrees well with the following 15. However, it is not clear why the word *šinīša* (twice, a second time) is here (if this is indeed the word). An alternative approach would be to read *irbi*, from *irbum*, income. In this case, the scribe detaches the *igi* of the income twice, which can be made consistent with the 2 sila$_3$ he harvested (line 3): the *igi* of 2 sila$_3$ twice—that is, the *igi* of 4—is 15.

Lines 6 and 7 seem to be easily read: "raise 15 to half (*ba-a*) of your field (*qa-ni-ka*) and 7 30 you see. The *igi* of 7 30 you detach and 8 you see". The same can be said of line R1: “16 you see. Raise 16 to 1”. Then, in lines R2 and R3, the result of the last raising is taken and raised to some other element to produce 8: “Raise 16 to … the *bu-ur* of your field (a.ša$_3$-*ka*) and 8 you see”. In fact, 1 *burum* = 30,0 sar$_s$, and 16 raised to 30,0 equals 8,0,0. The last part of line R3 seems to be a partial answer to the problem: “8 (is) your [x] (8 [x]-*ka*)”.

Finally, it is worth noticing the possible metrological bias of this tablet: cubits, *burum*, sila$_3$ and reed may be units present in it.

4.9 IM53965

4.9.1 References, Physical Characteristics and Contents

The tablet was published initially by Baqir (1951). von Soden (1952) offered a new reading for lines 3 to 5. However, neither Baqir nor von Soden presented a satisfactory mathematical interpretation. Bruins (1953a) proposed a new interpretation for the two lines on the upper edge and an organic mathematical explanation for the problem, which he presented in symbolic algebra. In what follows, I used Baqir’s original reading together with the improvements made by von Soden (only lines 3–5) and Bruins (lines 10 and 11, on the edge). The tablet was found in room 252, during the fourth season of work, in 1949. It measures 7.5 × 5.5 × 2.4 cm.

4.9.2 Transliteration and Transcription

Obverse

(1) *šum-ma ki-a-am i-ša-al-[ka*$^?$ *um-ma šu-u*$_2$*-ma*]
šumma kīam išâlka umma šū-ma
(2) *qa-na-am el-qe*$_2$$^?$*-a-ma* [*mi-in-da-su*]
qanâm elqeam-ma mindassu
(3) *u*$_2$*-ul i-de-e*!*-ma*! *šu*!*-*[*ši*] *ši-da-am al-li-ik*
ul īde-ma. šūši šiddam allik.
(4) *am-ma-at aḫ-ṣu*$_2$*-ub*$_2$*-šu-ma ša-la-ši*! *pu-ta-am*
ammat aḫṣubšū-ma šalāši pūtam
(5) [*a*]*l-li-ik* a.ša$_3$ 4.10 *ši-di u*$_3$! *pu-tu*
allik. eqlum 4.10. *šiddī u pūtu*
(6) [*ki*] *ma-ṣi*$_2$ *at-ta i-na e-pe*$_2$*-ši-ka*
kī maṣi. atta ina epēšika
(7) [*i*]*-gi* 30 *pu-ti-ka pu-ṭu*$_2$*-ur-ma*
igi 30 *pūtika puṭur-ma*
(8) 2 *ta-mar* 2 *a-na* 4.10 a.ša$_3$*-ka i-ši-ma*
2 *tammar.* 2 *ana* 4.10 *eqelka išī-ma*

Reverse

(R1) 8.20 *ta-mar* 8.20 *re-eš$_{15}$-ka li-ki-il*
8.20 *tammar.* 8.20 *rēška likīl.*
(R2) *na-as$_2$-ḫi-ir-ma am-ma-at ša ta-aḫ-ṣu$_2$-bu*
nasḫir-ma. ammat ša taḫṣubu
(R3) *a-na* 30 *pu-ti-ka i-ši-ma* 2.30 *ta-mar*
ana 30 *pūtika išī-ma* 2.30 *tammar.*
(R4) 2.30 *šu-ta-ki-il-ma* 6.15 *ta-mar*
2.30 *šutākil-ma* 6.15 *tammar.*
(R5) [6.15 *a*]*-na* 8.20 *i-ši-ma* 8.26.15 *ta-mar*
6.15 *ana* 8.20 *išī-ma* 8.26.15 *tammar.*
(R6) 8.26.15 *mi-na-am* ib$_2$.si.e 22.30 ib$_2$.si.e
8.26.15 *mīnam* ib$_2$.si.e 22.30 ib$_2$.si.e
(R7) [22].30 *me-eḫ-ra-am šu-ku-un* 2.30
22.30 *meḫram šukun.* 2.30
(R8) [*a*]*-na iš-te-en ṣi$_2$-ib i-na iš-te-en*
ana ištēn ṣib, ina ištēn
(R9) [*ḫu*]*-ru-iṣ iš-te-en* 25 [*iš*]*-te-en* 20
ḫuriṣ. ištēn 25 *ištēn* 20.

Edge

(E10) [2]5 1 *na-šu-ra-am*
25 1 […]
(E11) 20 2 *na*$^?$*-ak*$^?$*-si*$^?$*-id*$^?$
20 2 […]

Side

(S12) 25 *ši-du-um* [… xx]
25 *šiddum*
(S13) 10$^?$ *pu-tu-u*[*m* …]
10 *pūtum* […]
(S14) [. . . . . .] *i*$^?$ [. . . . . .]

4.9.3 Philological Commentary

In the edge, *na-šu-ra-am* and *na*$^?$*-ak*$^?$*-si*$^?$*-id*$^?$ are unidentified forms. Thus, they are not listed in the vocabulary.

In line 2, von Soden reads *el-qe$_2$-e*!*-ma*; in line E10, von Soden reads 25(?) 1! *qa*!-[*na*] *šu-ra-am.*

Baqir's original reading differed from the one presented here in the following lines:

(3) *u$_2$-ul i-de ki*$^?$ *ma-ṣu$_2$*$^?$ *ši-da-am al-li-ik*
(4) *am-ma-at aḫ-ṣu$_2$-ub$_2$-šu-ma ša-la-šu pu-ta-am*
(5) [*a*]*l-li-ik* a.ša$_3$ 4 10 *ši-di-im pu-ti-im*$^?$

Edge

(E10) [2]5 *a-na šu-ra*[?]*-am* [x …]
(E11) […] *ša* […]

4.9.4 Translation

[1]If (someone) asks you thus, (saying) this: [2,3]I took a reed and I do not know its size. I went sixty (times) the length. [4,5,6]I broke off a cubit from it, and I went thirty times the width. The area is 4,10. How much are my length and the width? You, in your doing, [7]detach the *igi* of 0;30, your width, and [8]you see 2. Raise 2 to 4,10, your area, and [R1]you see 8,20. May 8,20 hold your head. [R2]Return. The cubit you broke off [R3]raise to 0;30, your width, and you see 2;30. [R4]Cause 2;30 to combine and you see 6;15. [R5]Raise (sic. Should be "add", as Baqir (1951) had already noticed.) 6;15 to 8,20, and you see 8,26;15. [R6]What does 8,26;15 make equal? It makes 22;30 equal. [R7, R8, R9]Place 22;30, the copy. Add 2;30 to one, cut off from the other. One is 25, the other is 20. [E10]25, the first, has to be broken(?). [E11]20, the second, has to be cut off(?). [S12]25 the length … [S13]10 the width … [S14] …

4.9.5 Mathematical Commentary

The problem deals with the unknown dimensions of a rectangle of area 4,10 (line 5) which is measured with a reed of unknown length (lines 2 and 3). It is known, however, that the length of the field equals sixty times (1,0 in sexagesimal notation) the reed (line 3) and that the width equals 30 times what is left from the reed after one cubit is broken off from it (lines 4 and 5). This situation is represented below in Fig. 4.14a.

The rectangle in Fig. 4.14a has the length equal to sixty times the reed. Its width is thirty times the reed. The rectangle is divided into two vertical, rectangular strips. The right one has the width equal to thirty times a cubit, that is to say, 30 times 0;5 nindan, which is equal to 2;30 nindan. As a result, the left strip is the rectangle of the problem, with area 4,10 and width equal to thirty times the reduced reed (i.e., after a cubit is broken off from it).

In order to solve the problem, the scribe starts by calculating 2, the *igi* of 0;30 (lines 7 and 8), which is the ratio of the horizontal scaling that transforms the figure into a square (Fig. 4.14b).

In this new square, the leftmost region has area equal to 4,10 raised to 2, that is to say, 8,20 (lines 8 and R1). The other two regions have width equal to 2;30 nindan (line R3). This leads us to a problem about a square, as represented in Fig. 4.15a, where a rectangle of known width (twice the value 2;30) is added to an unknown square, producing a rectangle of area 8,20.[13]

[13]The passage from Figs. 4.14b to 4.15a is the same as the passage from Fig. 2.2b to a, that is to say, it corresponds to the reduction of one case of problems about squares.

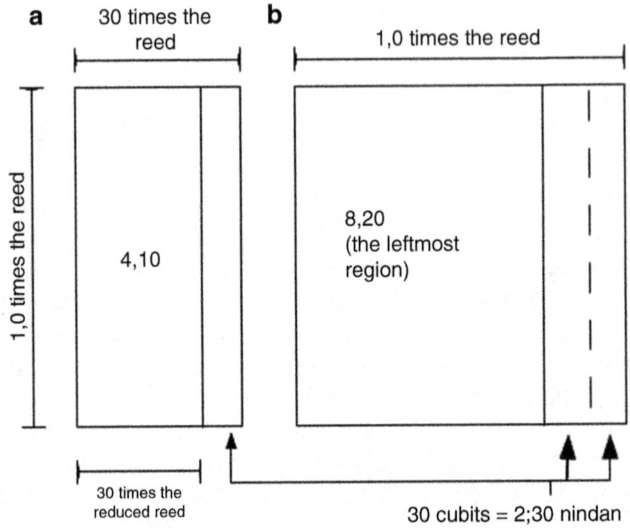

Fig. 4.14 (**a, b**) The statement of the problem and a scaling in IM53965

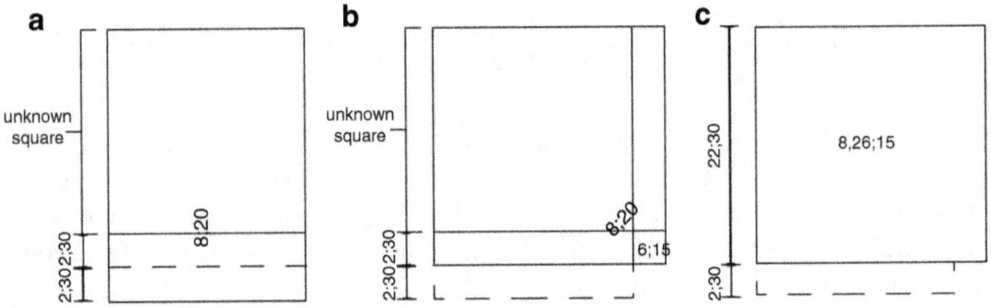

Fig. 4.15 (**a–c**) The problem about a square solved in IM53965

After cutting and pasting one of the smaller rectangles of width 2;30, an L-shaped region is formed, into which a square of area 6;15 (2;30 raised to 2;30, as in line R4) fits. This is shown in Fig. 4.15b. Thus the bigger square in Fig. 4.15c has area equal to the sum of 8,20 and 6;15, i.e., 8,26;15 (line R5). The scribe then asks what the side of this square is, and the answer is 22;30 (line R6).

The previous result is going to be used twice, and that is why a copy of it is placed (line R7). Now, 22;30 plus 2;30 results 25 (lines R7 to R9), which is the length of the rectangle of area 8,20 in Fig. 4.15a. On the other hand, 22;30 minus 2;30 results in 20 (lines R8 and R9), which can be thought of as the width of the same rectangle.

Finally, the horizontal scaling must be reversed. In doing this, the length 25 of the original rectangle in Fig. 4.14a is not affected (lines E10 and S12). The width of the rectangle, on the other hand, is 20 raised to 0;30, that is to say, 10 (lines E11 and S13).

4.10 IM54559

4.10.1 References, Physical Characteristics and Contents

The tablet was originally published by Baqir (1951). Unfortunately, part of the statement of the problem is destroyed, and reconstruction is possible only on a speculative basis. von Soden (1952) brought some amendments to the text, as well as a possible completion to the missing part of the statement. Bruins (1953a) proposed a different completion. Both von Soden's and Bruins's suggestions will be commented on below. The tablet was found in room 256,[14] during the fourth season of work, in 1949. It measures 8.2×6.2×2.3 cm.

4.10.2 Transliteration and Transcription

Obverse

(1) *šum-ma ki-a-am i-ša-al um-ma šu-u$_2$-ma*
šumma kīam išâl umma šū-ma
(2) *ši-ni-ip* uš sag.ki *a-na* [...]
šinip {šiddim pūtum/šiddim pūtim}ana [...]
(3) *a-na* sag.ki *u$_2$-ṣi$_2$-[ib]* a.ša$_3$ 20 uš sag.ki *mi*$^?$
ana pūtim uṣib. eqlum 20. *šiddum pūtum mi*$^?$
(4) [*mi*]*-nu-um at-ta i-na e-pe$_2$-ši-ka*
mīnum. atta ina epēšika
(5) ša.na.bi *a-na* 10 *i-ši-ma* 6.40 *ta-*[*mar*]
šinipêtim ana 10 *išī-ma* 6.40 *tammar.*
(6) *na-as$_2$-ḫi-ir* 10 *a-na* 1 *i-ši-ma* 10 *ta-mar*
nasḫir. 10 *ana* 1 *išī-ma* 10 *tammar.*
(7) 10 *e-li* 6.40 *mi-na-am wa-ta-ar*
10 *eli* 6.40 *mīnam watar.*
(8) 3.20 *wa-ta-ar na-as$_2$-ḫi-ir-ma* 3.20 *ḫi-pe$_2$-ma*
3.20 *watar. nasḫir-ma.* 3.20 *ḫipē-ma*
(9) 1.40 *ta-mar* 1.40 *šu-ta-ki-il-ma*
1.40 *tammar.* 1.40 *šutākil-ma*
(10) 2.46.40 *i-li* 2.46.40
2.46.40 *illi.* 2.46.40

[14] However, Hussein (2009, 92) states that this tablet also comes from room 252.

Edge

(E1) *a-na* 13.20 taḫ$_2$-*im-ma*
ana 13.20 *uṣim-ma*
(E2) 13.22.46.40 *i-li*
13.22.46.40 *illi.*

Reverse

(R1) 13.22.46.40 *mi-na-am* ib$_2$.si.[e]
13.22.46.40 *mīnam* ib$_2$.si.e
(R2) 28.20 ib$_2$.si.e 28.20 *me-eḫ-ra-am*
28.20 ib$_2$.si.e 28.20 *meḫram.*
(R3) 1.40 *ša tu-*[*uš*]*-ta-ki-lu a-na iš-te-en*
1.40 *ša tuštākilu ana išten*
(R4) *ṣi$_2$-ib i-na iš-te-en ḫu-ru-uṣ$_4$*
ṣib, ina išten ḫuruṣ.
(R5) [*i*]*š-te-en* 30 *i-li iš-te-en* 26.40
išten 30 *illi, išten* 26.40.
(R6) *n*[*a-as$_2$-ḫi-ir*]*-ma* 40 uš 30 [sag.ki]
nasḫir-ma 40 *šiddum.* 30 *pūtum.*

4.10.3 Philological Commentary

Table 4.8 summarises the divergent readings.

The second half of line 2 is too damaged to enable a reading of the signs. Thus, the sequence uš sag.ki may be read as two separate elements—namely, length and width—or just one composite element, as attested in other mathematical sources, the length–width, which refers to a rectangle. One should notice that length–width is still a composite of two words. A theoretical, composite logogram uš.sag.ki, corresponding to one word and meaning rectangle, is not accepted either in the CAD or in the AHw. Thureau-Dangin had used it in publications earlier to his TMB, but in the TMB he stated that it would be preferable to read uš sag.ki as two words, *šiddum-pūtum* (TMB 226). I will be back to this in the mathematical commentary.

Table 4.8 Divergent editorial readings in IM54559

Line	Sign(s)	Divergences
2	*a-na* […]	Read *a-na* 3[?] [*aš-ši-ma* uš uš] by von Soden (1952)
3	*a-na* sag.ki	Read *a-na* sag.ki 10! by von Soden (1952)
3	*u$_2$-ṣi$_2$-*[*ib*]	Read *u$_2$-ṣi$_2$*(?)-x by Baqir (1951)
3	*mi*?	Not read by Baqir (1951)
E1	taḫ$_2$-*im-ma*	Read *ṣi$_2$-im-ma* by Baqir (1951)
R5	26.40	Read 26.40 [2/3 uš] by von Soden (1952)

In line 3, the [*ib*] of u_2-*ṣi*$_2$-[*ib*] is not visible, as the tablet seems to be damaged at this point too. In Baqir's copy of the tablet, the last sign of line 3 reminds us of a *mi* or perhaps of a symbol for "10" followed by "*i*". Coincidentally, a *mi* is restored as the first sign of line 4, so there is the possibility that *mi-nu-um* is split in two parts, but this cannot be confirmed without direct access to the tablet.

As regards the transcription of ša.na.bi (line 5), see the comment on the entry *šinipum* in the vocabulary.

In line E1, *uṣim-ma* is written with a logogram and a phonetic complement: taḫ$_2$-*im-ma.* The verb is usually written with taḫ, but taḫ$_2$ seems to be a common writing in the analysed tablets, as this sign appears also in IM52301.

In other places, *meḫram* is followed by one of the imperatives *idi* (IM52301) or *šukun* (IM53965). Here, we can only read the accusative *meḫram*, for the remaining of line R2 is heavily damaged. This was also noticed by Baqir (1951).

4.10.4 Translation

[1]If (someone) asks thus, (saying) this: [2]two thirds of {the length the width/the rectangle} to [...] [3,4]to the width I added. The area is 20,0. What are the length, the width? You, in your doing, [5]raise two thirds to 10, and you see 6;40. [6]Return. Raise 10 to 1, and you see 10. [7]What does 10 go beyond 6;40? [8]It goes 3;20. Return. Halve 3;20, and [9]you see 1;40. Cause 1;40 to combine, and [10,E1]2;46,40 comes up. Add 2;46,40 to 13,20, and [E2]13,22;46,40 comes up. [R1]What does 13,22;46,40 make equal? [R2]It makes 28;20 equal. {Write down/Place} 28;20, a copy. [R3,R4]Add to one the 1;40 that you caused to combine. Cut off from the other. [R5]One: 30 comes up. Other: 26;40. [R6]Return, and the length is 40. The width is 30.

4.10.5 Mathematical Commentary

The problem deals with a rectangle. Part of its statement, however, is damaged, so that it is not possible to know exactly what the given data of the problem are. In line 2, there is a reference to two thirds either of the length (uš) or of the rectangle itself (uš sag.ki). The beginning of line 3 suggests that something is added to the width. The only piece of information that is readable is that the area is given as 20,0 (line 3). The problem seems to ask for the values of the length and the width (lines 3 and 4).

von Soden (1952) suggested that the statement of the problem reads as follows: I multiply 2/3 of the rectangle by 0;3, and then this gives the length. The length exceeds the width by 10. The area is 20,0. According to this interpretation, the value 30 that appears in lines R5 and R6 is the width of the rectangle; 40 is the length. The value 26;40 (line R5) would equal two thirds of 40, the length. Bruins (1953a) criticised von Soden's reconstruction on two bases. Firstly, that in order to read "two thirds of the rectangle", we should have a reference to the area of the rectangle written on the tablet (*ši-ni-ip* a.ša$_3$ uš sag.ki), but there is no such reference.

Secondly, that this statement would lead to a trivial problem: two thirds of the area raised to 0;3 equals two thirds of 20,0 raised to 0;3, that is to say, 13;20 raised to 0;3, namely, 40 the length. The width 30 would follow immediately, without being necessary to carry out the operations that are registered on the tablet.

Bruins (1953a) reconstructed the statement as follows: "Two thirds of the length. The width. On two times the length (3 × the width add 10)[15]; on the width, add 10. The area is 20". According to Bruins, this would be a "telegraphese" to say that "two times the length and three times the width have a difference of 10, irrespective of sign", leading to two possible cases, corresponding to the two possible values of the width, 30 and 26;40, as in the last lines of the tablet. To each of these widths, a corresponding length can be calculated: if 30 is the width, the length is 40; if 26;40 is the width, the length is 45. Thus,

- If two times the length is larger than three times the width by 10, we have width equal to 26;40 and length 45. Let us call this Case I.
- If three times the width is larger than two times the length by 10, we have width equal to 30 and length equal to 40. Let us call this Case II.

One interesting point in Bruins's interpretation is that it entails that the scribe would be able to solve simultaneously the two cases of problems about squares (represented in Chap. 2, respectively, by Figs. 2.1a and 2.1b). In order to bring evidence to his position, Bruins reported that IM31247 is a duplicate of this problem. Furthermore, while in IM54559 only the solution deriving from the width 30 is explicitly shown, in IM31247 the second solution "is regarded in more detail".

If, in fact, three times the width and two times the length have a difference, in any order, equal to 10, then one may deduce that the width itself and two thirds of the length have a difference of one third of 10 (that is to say, 3;20, as computed in lines 7 and 8). By raising each of these three mentioned values to the width, one obtains, respectively, a square with sides equal to the width, a rectangle with area equal to two thirds of the original rectangle (that is to say, 13,20, as is going to appear in line E1) and a rectangle with dimensions equal to the original width and to 3;20, the last figure having area equal to the difference of the areas of the first two, in any order they are taken. In other words, because of the two possible orders, two problems about squares are set:

Case I: a rectangle of area 13,20 is divided into a rectangle of width 3;20 and a square of unknown sides (the original unknown width), in Fig. 4.16a.

Case II: a square of unknown sides (the original unknown width) is divided into a rectangle of width 3;20 and a rectangle of area 13,20, as in Fig. 4.16b.

As already mentioned, these cases exemplify the two kinds of problems about squares solved by the cut-and-paste procedures explained in Chap. 2.

In order to solve these problems, the scribe performs the calculations described in the paragraph that follows. Figure 4.17a, b shows the corresponding geometrical interpretations for Case I, as in Fig. 4.16a. Similar figures can be drawn to represent Case II, but the calculations of the scribe are consistent with both cases.

---

[15] Which contained perhaps a mistyping. I propose "On two times the length (3 × the width) add 10; on the width, add 10".

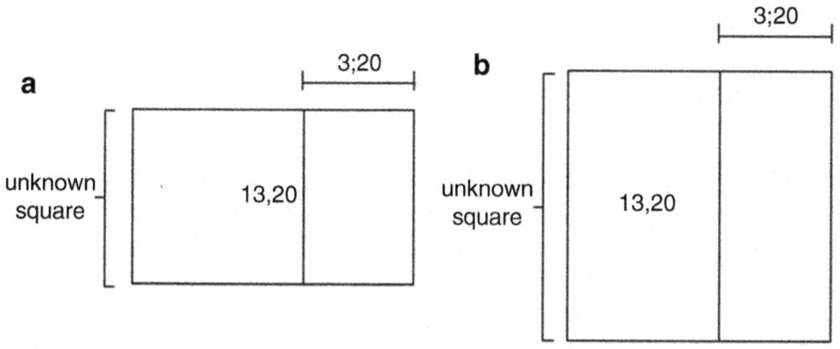

Fig. 4.16 (**a**, **b**) The two possible interpretations of the problem solved in IM54559

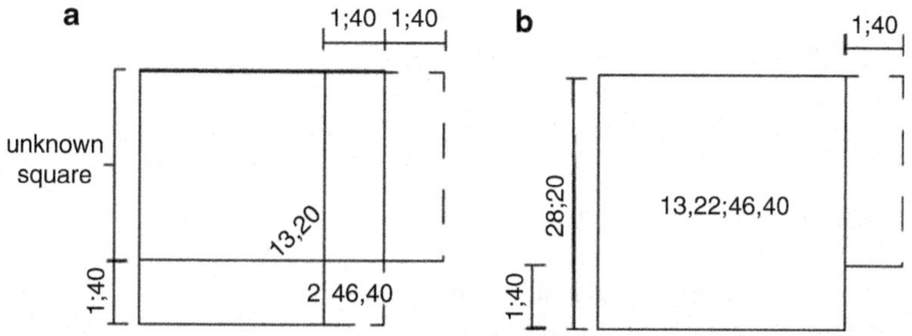

Fig. 4.17 (**a**, **b**) Solving the first case of the problem in IM54559

Firstly, the scribe halves 3;20, obtaining 1;40 (lines 8 and 9). This corresponds to dividing the rectangle of width 3;20 into two equal, thinner rectangles. One of these thinner rectangles is though of as having been moved to a new, horizontal position, producing an L-shaped region, as in Fig. 4.17a. In lines 9 and 10, the scribe causes 1;40 to combine, obtaining 2;46,40, which corresponds to the small square that fits the L-shaped region. Then the scribe adds to the area 13,20 the area of this small square, and the result is 13,22;46,40 (lines 10, E1 and E2), as in Fig. 4.17b. The rationale of doing this is to calculate the area of the bigger square that was formed when one of the rectangles of width 1;40 was placed in the horizontal position. The side 28;20 of this square is computed in lines R1 and R2.

In line R2, the scribe makes a copy of 28;20. He must then accumulate the first 28;20 with 1;40, obtaining 30 as the side of the original square in Fig. 4.16b. After that, he cuts off 1;40 from 28;20, which results in 26;40, the side of the original square in Fig. 4.16a. These operations are performed in lines R3, R4 and R5.

The solution of the first case produces 30 for the width (line R5) and, consequently, 40 is the length (line R6). As we do not know the conditions that were set up in the statement of the problem, it is not possible to know how 40 could be computed from 30. The solution of the second case, 26;40, might be another possible value for the width, with the corresponding length 45, which however does not appear explicitly in the text.

4.11 IM54464

4.11.1 References, Physical Characteristics and Contents

Baqir (1951) is the original publication, containing only the copy, the transliteration and some comments. von Soden (1952) offered important suggestions for improving the reading of some signs, which enabled him to be the first to publish a translation of the text. Bruins (1953a) accepted some of von Soden's suggestions, but disagreed in certain interpretive points (see, for instance, the philological commentary below). The tablet was found in room 252, during the fourth season of work, in 1949. It measures 7.5 × 5.5 × 2.5 cm.

4.11.2 Transliteration and Transcription

Obverse

(1) *šum-ma ki-a-am i-ša-al-ka um-ma šu-u*$_2$*-ma*
šumma kīam išâlka umma šū-ma
(2) *i-na ma-ḫi-ir* 1(ban$_2$) 5 sila$_3$ i$_3$.šaḫ 1(ban$_2$) i$_3$.giš
ina maḫir sūt 5 *qa naḫim sūt* {*ullum*/*šamnum*}.
(3) *ši-ni-ip ma-ḫi-ir na-ḫi-im u*$_3$ *ul-li-k*[*a*]
šinip maḫir naḫim u ullīka
(4) *wa-at-ri-im* gin$_2$ kug.babbar *na-ši-a-ku*
watrim. šiqil kaspam našiāku.
(5) i$_3$.giš *u*$_3$ i$_3$.šaḫ *ša-*[*ma-am*] *at-ta i-na e-pe*$_2$*-ši-ka*
{*ullam*/*šamnam*} *u naḫam šāmam. atta ina epēšika*
(6) *i-gi* 15 *pu-ṭu*$_2$*-ur-ma* [4 *i-li*]
igi 15 *puṭur-ma* 4 *illi.*
(7) 4 *a-*[*na*] 1 *i-ši-i-ma* 4 *i-*[*li-a*]*-kum*
4 *ana* 1 *išī-ma* 4 *illiakkum.*
(8) *na-as*$_2$*-ḫi-ir-ma i-gi* 10 *pu-ṭu*$_2$*-ur-ma*
nasḫir-ma. igi 10 *puṭur-ma*
(9) 6 *i-li* 6 *a-na* 40 [*i-ši*]*-i-ma* 4 *i-l*[*i*]
6 *illi.* 6 *ana* 40 *išī-ma* 4 *illi.*

Lower edge

(E1) *na-as*$_2$*-ḫi-ir-ma ul-li-im wa-at-r*[*i*]*-im*
nasḫir-ma. ullim watrim.
(E2) *i-na* 1 gin$_2$ kug.babbar 6 še kug.babbar
ina 1 *šiqil kaspim* 6 *uṭṭet kaspam*

Reverse

(R1) *ḫu-ru-uṣ*$_4$*-ma* 58 *ša-pi*$_2$*-il-tum*
ḫuruṣ-ma 58 *šapiltum.*
(R2) *na-as*$_2$*-ḫi-ir-ma* 4 *u*$_3$ 4 *ku-mu-ur-ma*
nasḫir. 4 *u* 4 *kumur-ma*
(R3) 8 *i-li* 8 *ša i-li-kum i-gi-šu pu-ṭ[u*$_2$*-ur-ma]*
8 *illi.*8 *ša illikum igišu puṭur-ma*
(R4) 7.30 *i-li* 7.30 *ša i-[li-a-kum]*
7.30 *illi.* 7.30 *ša illiakkum*
(R5) *a-na* 58 kug.babbar [*i-ši-i-ma* 7].15 [*i-li*]
ana 58 *kaspim išī-ma* 7.15 *illi.*
(R6) 7.15 *ša i-li-kum a-na* 4 *i-ši-i-ma*
7.15 *ša illikum ana* 4 *išī-ma*
(R7) 29 *i-na-di-na-ku-um na-as*$_2$*-ḫi-[ir]-ma*
29 *inaddinakkum. nasḫir-ma*
(R8) 7.15 *a-na* 4 *i-ši-i-ma* 2[9] *i-li*
7.15 *ana* 4 *išī-ma* 29 *illi.*
(R9) 6 še kug.babbar *ša i-na* kug.babbar *ta-su-ḫu*
6 *uṭṭet kaspam ša ina kaspim tassuḫu*

Upper edge

(e1) *a-na iš-te-en ṣi*$_2$*-ma*
ana ištēn ṣim-ma
(e2) *iš-te-en* 29 *iš-te-en* 31
ištēn 29. *ištēn* 31.

Side

(S1) *ša-am iš-te-en* 7.15
šām ištēn 7.15
(S2) *iš-te-en* 5.10
ištēn 5.10.

4.11.3 Philological Commentary

The divergent readings are summarised in Table 4.9.

Furthermore, it seems that approximately the same group of signs is repeated in the end of line 3 and in the middle of line E1. The three following different readings have been proposed for them:

- Baqir (1951) read *u*$_3$ *x*$^?$ *ul-li-an* (line 3) and *u*$_3^?$ *ul-li-an* (line E1), suggesting "excessive" and "above" as possible meanings, that he derived from Deimel's *Akkadisch-Sumerisches Glossar*. However, as already mentioned, Baqir refrained from translating the tablet.

Table 4.9 Divergent editorial readings in IM54464

Line	Sign(s)	Divergences
3, E1	*ul-li-k[a]*, *ul-li-im*	See commentary above
4	gin_2 kug.babbar	Read *x* 15$^?$ by Baqir (1951) and 2/3 gin_2 kug. babbar by von Soden (1952)
5	*ša-[ma-am]*	Read *ša* ... by Baqir (1951)
E1	The entire line	Read *na-as*$_2$*-ḫi-ir-ma* ki.lam *ul-li-im wa-at-r[i]-im* by von Soden (1952)
E2	gin_2 kug.babbar 6 še kug.babbar	Read su$^?$ *x x* 6,40 15 21$^?$ by Baqir (1951)
R4	*i-[li-a-kum]*	Read *i-[li-kum]* by Baqir (1951)
R5	kug.babbar	Read *x x* by Baqir (1951)
R9	6 še kug.babbar *ša i-na* kug.babbar	Read 6,40,15,21 *ša i-na* 15$^?$ 21$^?$ by Baqir (1951)
e1	*ṣi*$_2$*-ma*	Read *ṣi*$_2$*-ib* by Baqir (1951)

- von Soden (1952) read *ul-li-k[a]* (line 3) and *ul-li-[i]m*$^?$ (line E1) and commented that the Old Babylonian reading for ia_3.giš (that is to say, i_3.giš), at least in Ešnunna, is *ullum* rather than *ellum*.[16] As this solution seems to be quite consistent with the remaining of the text, it has been adopted here. Furthermore, the logogram i_3.giš can be rendered also as the Akkadian *šamnum*, oil. Thus, the logogram i_3.giš is transcribed here as {*ullum*/*šamnum*}, in order to give room for these two possibilities. Both Akkadian words are translated here simply as oil.
- Bruins (1953a) read 2 *ul ša an* (line 3), but did not propose any grammatical explanation for this. His translation runs as "Two thirds of the price of oil, and 2 is the excess of the oil price".

It is worth noticing the use of *inaddinakkum* (it gives to you) in line R7, instead of the usual *illiakum* (it comes to you).

4.11.4 Translation

[1]If (someone) asks you thus, (saying) this: [2]for the price of 1 ban_2 5 $sila_3$ of lard, 1 ban_2 of oil. [3,4]Two thirds of the price of lard and of your exceeding oil. I have brought one shekel of silver. [5]Buy oil and lard to me. You, in your doing, [6]detach the *igi* of 15, and 0;4 comes up. [7]Raise 0;4 to 1, and 0;4 comes up to you. [8]Return. Detach the *igi* of 10, and [9]0;6 comes up. Raise 0;6 to 0;40, and 0;4 comes up. [E1]Return. Of exceeding oil. [E2,R1]Cut off from 1 shekel of silver 6 grains of silver and the remainder is 0;58. [R2]Return. Accumulate 0;4 and 0;4, and [R3]0;8 comes up. Detach the *igi* of 0;8 that comes up to you, and [R4,R5] 7;30 comes up. Raise 7;30 that comes up to you to 0;58, the silver, and 7;15 comes up. [R6]Raise 7;15 that comes up to you to 0;4, and [R7]it gives 0;29 to you. Return. [R8]Raise 7;15 to 0;4, and 0;29 comes up. [R9,e1]You add 6 grains of silver that you removed from silver to one, and [e2]one is 0;29. The other is 0;31. [S1]Buy one 7;15, [S2]the other 5;10.

[16]The occurrences of this word are registered under *ellum* in the vocabulary.

4.11.5 Mathematical Commentary

In this problem, 1 shekel of silver must be used to buy lard and oil (lines 4 and 5), so that 6 grains more of silver are spent with oil. This last piece of information, however, is not explicitly given in the statement of the problem, but can be deduced from its solution, as we shall verify below.

We are informed that 15 sila$_3$ (1 ban$_2$ 5 sila$_3$) of lard are equivalent to 10 sila$_3$ (1 ban$_2$) of oil (line 2). It is also tacitly assumed that each of these quantities costs 1 shekel of silver. Lines 3 and 4 seem to reinforce the relation between the prices, mentioning "two thirds of the price of the lard" and "of your exceeding oil" but, as the last expression comes in the genitive, it is not possible to give a direct translation. Perhaps the scribe's intention is to say that "two thirds: the exchange of the lard and your exceeding oil", but this would entail the presence of two complements in the genitive (lard and exceeding oil) for the same status constructus (*maḫir*). von Soden (1952), on the other hand, joined the whole of lines 3 and 4 in only one sentence: "Als 2/3 des Gegenwerts des Schmalz' und deines(?) überschüssigen Feinöls, bringe ich 2/3 (sic) Sekel silver". However, this solution does not make clear why two thirds of the price of lard (instead of oil) are mentioned. Furthermore, it fails to account for other missing data. As von Soden himself recognises, in line 3 there should be perhaps some more room to accommodate the six grains more that are to be spent with oil (1952, 54, note 14).

Once 15 sila$_3$ of lard costs 1 shekel of silver, the scribe detaches the *igi* of 15, obtaining 0;4 (line 6). This value is raised to 1 shekel, thus giving the price of 1 sila$_3$ of lard (line 7). In lines 8 and 9, a similar, abbreviated computation gives the price of oil: the scribe detaches the *igi* of 10 and obtains 0;6 shekel for each sila$_3$ of oil. The final part of line 9 shows that two thirds (0;40) of the price of oil is the price of lard, and this may be related to the two thirds that appear in the beginning of line 3.

Line E1 is of difficult interpretation. von Soden (1952) joins it with line E2 and proposes the translation "Wende dich, und als Gegenwert (?) des überschüssigen Feinöls ziehe von 1 Sekel Silber 6 Korn Silber". This might indeed have been the scribe's intention, but this reading depends upon the introduction of the logogram ki.lam, but it seems that there is no available room for it. Anyway, lines E2 and R1 show that the scribe cuts off from 1 shekel of silver, an amount of six grains that supposedly must be spent with oil. As six grains equal 0;2 shekel, the remainder is 0;58 (line R1), which is to be spent equally with lard and oil. The next step is to accumulate 0;4 and 0;4, that is to say, to obtain 0;8 as twice the price of 1 sila$_3$ of lard (lines R2 and R3). The rationale behind this, in my opinion, is that if the price of lard was the double of its actual price, then 0;58 shekel of silver would be enough to buy only lard. That is why the scribe computes the *igi* of 0;8, concluding that it would be possible to buy 7;30 sila$_3$ of lard with 1 shekel (lines R3 and R4); then he raises this value to 0;58 shekel and obtains 7;15 sila$_3$ (line R5), which is the amount of lard that he will be able to buy. By raising 7;15 sila$_3$ of lard to its actual price, 0;4, the scribe obtains 0;29 shekel, which is indeed the half of 0;58 shekel of silver (lines R6 and R7). After a repetition of the raising of 7;15 to 0;4 (line R8), he adds to 0;29

shekel the 6 grains previously removed (line R9). Thus, 0;29 shekel is to be spent with lard and 0;31 shekel is to be spent with oil (lines e1 and e2). In line S1, the scribe writes the already known fact that 7;15 $sila_3$ of lard are to be bought. Finally, in line S2, without indicating the computation, he presents the amount of oil that is to be bought, 5;10 $sila_3$ of oil. We can complete this last step by noticing that the *igi* of the price of oil, that is to say, the *igi* of 0;6 is 10 $sila_3$ per shekel. Raising 0;31 shekel by this number, we obtain 5;10 $sila_3$ of oil.

4.12 IM54011

4.12.1 References, Physical Characteristics and Contents

The tablet contains two related problems about the construction of a brick wall. It was originally published by Baqir (1951), but due to its bad state of preservation only a copy and a partial transliteration were presented. von Soden (1952) brought a proposal of a complete transliteration, a translation and an interpretation, which however leads to a situation identified as highly improbable: the brick wall, according to von Soden's analysis of the text, would have its top thicker than its bottom. Bruins (1953a) proposed a different interpretation, but still relying on some of von Soden's readings. Robson (1999, 94–96) published a new complete transliteration, translation and mathematical commentary, placing the tablet in the more comprehensive context of earthworks and coefficient lists. The tablet was found in room 252, during the fourth season of work, in 1949. Its measures were not given explicitly by Baqir (1951) in his text; the accompanying photograph, however, suggests that the size of this tablet is not substantially different from the other nine that were published at the same time.

4.12.2 Transliteration and Transcription

Obverse

(1) [*šum-ma ki-a-am*] *i-ša-al-ka um-ma šu-u*$_2$*-ma*
šumma kīam išâlka umma šū-ma
(2) *pi*$_2$*-ti-iq-tum a-ša-al ši-du-um ši-ta am-ma-tim*
pitiqtum. ašal šiddum šitta ammātim
(3) *ru-up-šu-um* kuš$_3$ {*ḫi-pe*$_2$/*ḫe-pi*$_2$} *a-na e-le-num* [*ku-bu*]*-ur*
rupšum ammat {*ḫipe/ḫepi*} *ana elēnum kubur*
(4) *ni-ka-as*$_2$ *mu-lu-um e-pe*$_2$*-ru-ka u*$_3$ [*ṣa-bu-ka*]
nikkas mūlûm. eperuka u ṣābūka
(5) *u*$_2$*-ma-ka-lu-tum mi-nu-um at-ta i-na* [*e-pe*$_2$*-ši-ka*]
ūmakallūtum mīnum. atta ina epēšika
(6) *ši-ta am-ma-tim ru-up-ša-am u*$_3$ kuš$_3$ {*ḫi-pe*$_2$/*ḫe-pi*$_2$}

šitta ammātim rupšam u ammat {*ḫipe/ḫepi*}
(7) *ku-mu-ur ḫi-pe*$_2$*-e-ma* 6.15 *ḫe-pu-šu*
kumur. ḫipē-ma 6.15 *ḫepûšu.*
(8) [6].15 *a-na ni-ka-as*$_2$ *mu-li-im i-ši-*[*ma*]
6.15 *ana nikkas mūlîm išī-ma*
(9) [1]8.45 *ta-mar* 18.45 *a-na a-ša-*[*al ši-di-im*]
18.45 *tammar.* 18.45 *ana ašal šiddim*

Edge

(E1) *i-ši-ma* 3.[7.30] *ta-mar* 3.7.30 [*e-pe*$_2$*-ru-ka*]
išī-ma 3.7.30 *tammar.* 3.7.30 *eperuka.*
(E2) *na-as*$_2$*-ḫi-ir-ma* 3.45 *i-gi-gu-*[*ub-bi-ka*]
nasḫir-ma. 3.45 *igigubbika.*

Reverse

(R1) [*i*]*-gi* 3.45 *pu-ṭu*$_2$*-ur-ma a-na* [3.7.30]
igi 3.45 *puṭur-ma ana* 3.7.30
(R2) *e-pe*$_2$*-ri-ka i-ši-ma* 50 *ta-*[*mar* 50 *ṣa-bu-ka*]
eperika išī-ma. 50 *tammar.* 50 *ṣābūka.*
(R3) *šum-ma ki-a-am i-ša-al-ka um-ma šu-u*$_2$*-*[*ma*]
šumma kīam išâlka umma šū-ma
(R4) [*iš*]*-ka-ar a-wi-lim iš-te-en* [*ki*] *ma-*[*ṣi*$_2$]
iškar awīlim ištēn kī maṣi.
(R5) *at-ta i-na e-pe*$_2$*-ši-ka ši-ta* [*am-ma-tim*]
atta ina epēšika šitta ammātim
(R6) [*ru-up-ša*]*-am u*$_3$ kuš$_3$ {*ḫi-pe*$_2$*/ḫe-pi*$_2$} [*ku-mu-ur*]
rupšam u ammatam {*ḫipe/ḫepi*} *kumur.*
(R7) [*ḫi*]*-pe*$_2$*-ma* 6.15 *ḫe-pu-šu* [6.15] *a-na*
ḫipē-ma 6.15 *ḫepûšu.* 6.15 *ana*
(R8) [3 *mu*]*-li-im i-ši-ma* [18].45 *ta-mar*
3 *mūlîm išī-ma* 18.45 *tammar.*
(R9) [*i-gi* 18].45 *pu-ṭu*$_2$*-ur-ma a-na* 3.45 [*i-ši-i-ma*]
igi 18.45 *puṭur-ma. ana* 3.45 *išī-ma*
(R10) [12 *iš-ka-ar*]*-ka-ma*
12 *iškarka-ma.*

4.12.3 Philological Commentary

The above transliteration is almost entirely based on von Soden (1952) and Robson (1999), and as a result it is highly different from that originally published by Baqir (1951). Table 4.10 lists my own divergent readings in relation to von Soden and Robson.

It should be noticed that, in line E2, Robson proposes *i-gi-gu-[ub pi$_2$-ti-iq-tim]*, where von Soden writes *i-gi-gu-[ba-ka]*. In my opinion, it is more likely that the scribe wrote a third possibility, *i-gi-gu-[ub-bi-ka]*, which is a status constructus in the nominative, followed by a possessive suffix. There is no necessity for the scribe to make explicit that he is dealing with a brick coefficient, so he might simply omit *pi$_2$-ti-iq-tim*.

The expression kuš$_3$ *ḫi-pe$_2$* is problematic. Inserted in the middle of sentences in lines 3, 6 and R6, it seems to produce a broken syntax. Robson transcribes it as kuš$_3$ *ḫe-pi$_2$* and gives as translation "a broken cubit". In order to understand which form is dealt with here, let us first notice that it cannot be a stative, for that would still produce a broken syntax. More, who or what would be the subject of this stative, in the singular third masculine person? Yet, *ḫe-pi$_2$* can be interpreted as a form of the adjective *ḫepûm*. Of course, it is not the status rectus, for *ammatum* is feminine (as is shown by the expression *ammatum rabītum* (CAD A2, 74, s.v. *ammatum*)), so we should have *ḫepītam*. The status constructus can also be eliminated, on the same grounds of gender concordance. As Robson makes us notice, it "must be in the absolute state, as is common with units of measure" (1999, 96). Besides this possibility, in the transliteration, transcription and translation above, I also retained the reading of an imperative, *ḫi-pe$_2$*, to be interpreted as a kind of parenthesis in the sentence. If on the one hand, this makes the syntax less smooth, on the other it contemplates a very common form in mathematical texts. As already explained in Sect. 3.1, two different possible readings are inserted in curly braces and separated by a slash: {*ḫi-pe$_2$/ḫe-pi$_2$*}. In line 7, *ḫe-pu-šu* is apparently an infinitive (or maybe an adjective) followed by a possessive suffix.

For the expression *šitta ammātim*, see Sect. 2.4. In line 6, on the other had, *ammatim* may simply play the function of the direct object of *ku-mu-ur*.

In the end of line 3, *[ku-bu]-ur* is most likely a status absolutus. It could be a status constructus, if it was followed by a suffix or a genitive (*kuburšu* or *kubur pitiqtim*), but as there is no visible available room in the tablet for an additional sign, this is probably not the case.

The writings *ḫi-pe$_2$-e-ma* (line 7) and *i-ši-i-ma* (line R9) attest that under certain conditions, the particle *ma* causes a lengthening of the preceding vowel. In the analysed tablets, by the way, the writing *i-ši-i-ma* is relatively frequent.

4.12.4 Translation

[1]If (someone) asks you thus, (saying) this: [2,3]a brickwork. The length is a rope, the width is two cubits, {one broken cubit/one cubit --- halve it ---} is the thickness towards above and [4,5]the height is a *nikkassum*. What are your volume and your workers for 1 day? You, in your doing, [6,7]accumulate two cubits, the width, and {one broken cubit/one cubit --- halve it}. Halve it, and 0;6,15 is its halving. [8]Raise 0;6,15 to a *nikkassum*, the height, and [9,E1]you see 0;18,45. Raise 0;18,45 to a rope, the length, and you see 3;7,30. Your volume is 3;7,30. [E2]Return. Your coefficient is

Table 4.10 Divergent editorial readings in IM54011

Line	Sign(s)	Divergences
7	---	Insertion of [*ta-mar*] at the end of the line by Robson (1999)
E2	*i-gi-gu-[ub-bi-ka]*	Read *i-gi-gu-[ub pi*$_2$*-ti-iq-tim]* by Robson (1999)
R3	*šu-u*$_2$*-[ma]*	Read *šu-u*$_2$ by Robson (1999)
R7	*ḫe-pu-šu* [6.15]	Read *ḫe-pu-šu* <*ta-mar*> [6.15] by Robson (1999)
2,6, R5	*am-ma-tim*	Read *qa-ta-tim* by von Soden (1952)
3,6	kuš$_3$ *ḫi-pe*$_2$	Read *ammat* 40 *qanûm* by von Soden (1952)
3	*[ku-bu]-ur*	Read *wu-tu-ur* by von Soden (1952)
E2	*i-gi-gu-[ub-bi-ka]*	Read *i-gi-gu-[ba-ka]* by von Soden (1952)
R1	[3.7.30]	Read [3].7.2[0] von Soden (1952)
R6	The whole line	Read [*ru-up-ša-am u*$_3$ *ammat* 40 *qanûm*? *pi*$_2$*-ti*?[*-iq*?*-tam*?] *k*[*u-mu-ur*] by von Soden (1952). Due to an obvious typo, the first bracket is not closed
R7	[6.15] *a-na*	Read [6.1]5 *a-na-ni-ka-as*$_2$] by von Soden (1952). As in the previous case, the last bracket does not have a pair
R8	[3 *mu*]	Read [*mu*] by von Soden (1952)
R9	*pu-ṭu*$_2$*-ur-ma a-na*	Read *pu-ṭu*$_2$*-ur-ma* 3.12 *a-na* by von Soden (1952)

0;3,45. [R1,R2]Detach the *igi* of 0;3,45, and raise to 3;7,30, your volume. You see 50. Your workers are 50. [R3]If (someone) asks you thus, (saying) this: [R4]how much is the work of one man? [R5,R6]You, in your doing, accumulate two cubits, the width, and {one broken cubit/one cubit --- halve it}. [R7,R8]Halve it, and its halving is 0;6,15. Raise 0;6,15 to 3, the height, and you see 0;18,45. [R9]Detach the *igi* of 0;18,45. Raise to 0;3,45, and [R10]your work is 0;12.

4.12.5 Mathematical Commentary

This problem deals with the construction of a brick wall. The statement of the problem informs us, in lines 2–4, that the wall has length equal to one rope (ten nindan). The cross section of the wall is a trapezium with bases 2 cubits (the width of the wall) and half a cubit (the thickness towards above). Its height is one *nikkassum* (3 cubits). The problem requires the calculation of the volume of the wall and the number of men necessary to build it in 1 day (line 5).

The scribe starts the solution by accumulating two cubits and half a cubit, that is to say, the dimensions of the bases of the cross section of the wall, and by halving the result. There are two ways to interpret what the scribe does.

Traditionally, it has been supposed that the scribe converts these measures to nindan: 2 cubits equals 0;10 nindan; half a cubit equals 0;2,30 nindan; their accumulation equals 0;12,30 nindan, so its half is 0;6,15 nindan (lines 6 and 7). This number is raised to 3, the cubit equivalent of the height of one *nikkassum*, giving 0;18,45 as the area (in mixed units nindan × cubit) of the cross section of the wall (lines 8 and 9). Next, the scribe raises this area to the length of the wall, namely 10 nindan (a rope), obtaining 3;7,30 sar_v, which is the volume and so the answer to the first part of the question (lines 9 and E1).

A second interpretation assumes that the scribe converts the measures to abstract numbers, with which it would be possible to perform arithmetic operations. This is in line with what Proust proposes as part of her interpretation to the mathematical school tablets of Nippur (Proust 2007). Although from the point of view of the numerical values that are involved, this interpretation is equal to the previous one, from the point of view of the cognitive tools the scribes used it is quite different. Using initially a table of lengths, the scribe converts 2 cubits to the abstract number 10, that is to say, 10 without either measuring units or a fixed order of magnitude. Half a cubit is converted to 2.30. Their accumulation is a pure arithmetical operation, resulting in 12.30. The scribe halves it, and 6.15 is the result. Using again a metrological table, 1 *nikkassum* is converted to 3. The scribe raises 6.15 to 3, both "abstract" numbers, and the result is 18.45. A rope is converted to 10. By raising 18.45 to 10, the scribe arrives at the desired result, 3.7.30.

With the aid of the coefficient 0;3,45, the volume of wall one man is capable of building in 1 day, the scribe calculates the number of men required to build this wall in just 1 day. He takes the *igi* of 0;3,45 and raises it to the volume 3;7,30 sar_v, by which he concludes that 50 men build this wall in 1 day (lines E2, R1 and R2).

Line R3 opens a new problem, although related to the one just discussed. We are now required to compute the length of wall one man builds in 1 day (line R4). In lines R5 to R8, the area of the cross section is computed again, following exactly what had been done previously. The coefficient 0;3,45 is taken again too (line R9). As this coefficient gives a volume one man builds in 1 day, the scribe raises the *igi* of the area of the cross section to it, obtaining 0;12 nindan as the result, which is the length of wall built by one man (lines R9 and R10). This is obviously consistent with the results of the first part: if one man builds 0;12 nindan of wall, then 50 men build 10 nindan, that is to say, one rope.

Chapter 5

On Old Babylonian Mathematics and Its History: A Contribution to a Geography of Mathematical Practices

5.1 A Few Preliminaries

Mathematical tablets come from archaeological sites that represent a wide spectrum of places and periods. There are mathematical tablets from the cities of Larsa, Ur and Uruk, in the extreme south of Mesopotamia; Isin, Nippur, Babylon and Kiš, going to the north; Susa, in Elam; Šaduppûm, Ešnunna, Zaralulu, Nerebtum and Me-Turan, in the Diyala region; Assur and Niniveh, in the north; Mari and Terqa, going to the west; Ebla and Hazor, almost on the Mediterranean coast; Ugarit, on the very coast of the Mediterranean and Amarna, in Egypt. These tablets range from the late fourth millennium to the Seleucid period. However, the vast majority of mathematical tablets date to the Old Babylonian period.[1]

During the Old Babylonian period, much, but not all, of the mathematical practices seemed to be linked to the environment of scribal education. A great deal of the evidence consists of school exercise tablets and sometimes teacher's models made for the pupils' profit. The Old Babylonian scribal schools, however, were not edifices built especially for the purpose; the education of the apprentices was in some cases performed at the residences of their masters (Tinney 1998, 41). Mathematics, lexical lists, music and short Sumerian compositions were the subjects to be studied by young scribes (Sjöberg 1976). The mathematical curriculum started with the study of tables of multiplication, squared numbers, square and cube roots and reciprocals; it continued with the study of metrological material; and, for those that went to the advanced level, the curriculum seemed to be completed by the

[1] Comprehensive lists of mathematical tablets, their contents, provenance (when known) and publication references are given by Nemet-Nejat (1993, 103–148, 251–290) and Robson (2008, 299–344). See Gonçalves (forthcoming) for a list covering the Diyala.

C. Gonçalves, *Mathematical Tablets from Tell Harmal*, Sources and Studies in the History of Mathematics and Physical Sciences,
DOI 10.1007/978-3-319-22524-1_5

study of problems involving utilitarian as well as more abstract, mathematical matters, as exemplified by the tablets examined in Chap. 4.[2]

Is Mesopotamian mathematics, or cuneiform mathematics, or specifically Old Babylonian mathematics, appropriately called *mathematics*? As is well known, the term *mathematics* is of Greek coinage and indicated originally a group of subjects recognised, not always with perfect agreement, as mathematical, that is to say, whose learning could be reached only as the result of active and conscious studying. Later thinkers—philosophers or mathematicians, sometimes both—accepted the label *mathematical* for an always changing group of disciplines, but no Old Babylonian scribe ever did it, nor could, having lived at least a thousand years before the oldest of the ancient Greek mathematicians.[3] There is not either any word for designating mathematics in the Mesopotamian sources.[4] Yet, grouping certain cuneiform tablets into a corpus and calling this corpus *Mesopotamian mathematics*, or the like, has been thought to be possible on two grounds. Firstly, Old Babylonian culture conferred some degree of unity to this corpus. It was produced mainly in scribal schools, its language was highly standardised (as it is possible to verify, by reading the tablets presented in Chap. 4), its contents were very well delineated and differentiated from other genres of text[5] and there are, in Old Babylonian texts about school life, the eduba texts, explicit references to the study of what this corpus approximately contains. Frequently repeated examples to sustain the last affirmation are the following, which I quote from Sjöberg (1976, 167–168):

- In the *Dialogue 1*, we read "you may recite the multiplication table, but you do not know it perfectly; you may solve inverted numbers, (but) you cannot …"
- In *Enkimansum and Girini-isag*, "go to divide a field but you won't be able to divide it, go to delimit a field but you won't be able to hold the tape and the measuring rod, the pegs of your field you won't drive in, you are not able to figure out the sense".
- In *Examination Text A*, "Do you know multiplication, reciprocals, coefficients, balancing of accounts, administrative accounting, how to make all kinds of pay allotments, divide property and delimit shares of fields?" [6]

The second reason to call Old Babylonian mathematics *mathematics* is a practical one: as Old Babylonian scribes did not leave a general term to systematically

[2] For the mathematical curriculum, see Proust (2007) and Robson (2009, 199–226).

[3] The original Greek meaning is taken here from the book of *Definitions*, by the pseudo-Heron (Heron-Heiberg 1912). For the changing character of the mathematical disciplines, although not in Mesopotamia, see my Gonçalves (2012).

[4] No word for "number" either.

[5] An interesting example of this is given by the so-called anthology texts that mixed different mathematical topics but never non-mathematical matters.

[6] *Examination Text A* is an embarrassing example to understand Old Babylonian school, for its oldest copy dates to approximately 900 B.C.E. However, there is no doubt that the composition on the whole reflects the Old Babylonian eduba (Sjöberg 1974, 10; 1976, 160). In particular, the sentence reproduced here is among those that have "parallels in Old Babylonian texts dealing with eduba" (Sjöberg 1976, 160).

and clearly refer to these tablets, [7] and as the texts are indeed akin to what *we* call mathematics, it is very handy to speak of Mesopotamian, or cuneiform, or Old Babylonian mathematics. [8]

This leads us to the problem of speaking of *mathematicians* in the Mesopotamian setting. Once one resigns themselves to using *mathematics* as a label for a specific Old Babylonian practice, it would make sense to call its practitioners *mathematicians*. However, the field usually refers to the people who wrote cuneiform mathematics as *scribes* or *students* of the scribal profession. [9] Firstly, because this makes clear that accepting *mathematics* as an isolated term, in part for practical reasons, is quite different from accepting the whole semantic field represented by *mathematics*, *mathematical* and *mathematicians*, clearly extraneous to Old Babylonian culture—the ancient scribes and students were not pursuing an academic discipline devoted to the finding of new results and the development of theory. In the second place, calling the authors of these tablets *scribes* or *students* makes sense because mathematical tablets were in fact written either by scribes or by students.[10] In many occasions, it is possible to clearly identify that a tablet was written by a student. In others, as the tablets studied in Chap. 4, it is more difficult to assert whether they are model problems written by a teacher or the result of the efforts of an advanced pupil. For the sake of simplicity, in cases like this, both Assyriologists and historians of mathematics tend to use the term *scribe*.

Being now clear in what sense we can speak of Old Babylonian mathematics (and its practitioners, scribes and students), one may inquire if it constitutes a chapter of the *history of mathematics*. The main problem here is the one of conceptually establishing the discipline of the history of mathematics. The extension of this problem can be better evaluated if we ask the following question: are Old Babylonian mathematics, Egyptian mathematics, European Medieval mathematics, modern twenty-first century mathematics, to mention just these cases, parts of the one and same human practice, for which it is possible to write a *history*, in the sense of a series of transformations through which it would have passed over a span of thousands of years? If this were the case, then we would be justified in postulating a continuity from Old Babylonian mathematics up to the present mathematical practices, which seems to be a rather improbable historiographical project. Besides, we should be able to account for the many possible contacts and influences between

---

[7] As Robson (2008, 289) reports, "technical terms were used by particular individuals in specific contexts". None of these terms, however, was systematically used to define an Old Babylonian mathematical discipline or disciplines. See also Nemet-Nejat (1993, 6–10) for some of these terms.

[8] It could not go without noticing that the term *Mesopotamia* is also of Greek coinage. The words *mathematics* and *Mesopotamia* are, thus, witnesses of our adherence to a Greek-oriented tradition.

[9] Nemet-Nejat presents expressions that in lexical texts referred to some sort of mathematically oriented professionals: "man of (the) accounting board(?), man of (the) abacus(?), man of stone(s/weights), man of clay stone(?)" (1993, 5).

[10] Akkadian *ṭupšarrum* is the Old Babylonian word that refers to the person who writes tablets, a *scribe*.

Mesopotamia, Egypt and Greece, which again is quite controversial. Finally, we should integrate in such a large synthesis the many local mathematics, as those of the American indigenous peoples, the mathematics of commercial activities and so on. Due to all these obstacles, in my opinion, it is much more reasonable to consider the discipline of the history of mathematics as a convergence point for historians whose interests were led to one of those very different *mathematical traditions*. This is the only way, I think, to treat cuneiform mathematical tablets with respect for their own terms—in the literal and figurative sense—and this is the basis for the historiographical stance defended in this work.

In order to illustrate the last sentence of the previous paragraph, it may be worth emphasising the following. Once the 12 cuneiform tablets studied here are part of the specific Old Babylonian mathematical tradition, it is not historiographically productive to analyse them through algebraic symbolic tools, which are typical of other lineages of mathematical practices. Instead, if we use their own words and expressions for the arithmetical operations, if we recognise the techniques scribes used (as cut-and-paste geometry and scaling of figures), if we legitimate conceptual differences as that between abstract numbers and numbers accompanied by units of measure and if we keep in mind that each tablet is an ancient clay object coming most likely from an educational environment, then we are able to describe the practices and beliefs Old Babylonian individuals had as regards numbers, shapes, measures and the world around them. In other words, we are able to know more about their way of thinking and about what they believed were the main characteristics of the world they lived in.[11]

As a result, the tablets presented in this work are seen rather as a part of Old Babylonian history than of a history of mathematics in an idealised sense. Instead of asking what these tablets have to do with the mathematics of later periods, or how Old Babylonian mathematics fits an imaginary world development of mathematics, we can ask why Old Babylonian scribes wrote mathematics and why Old Babylonian mathematics took that specific form. Finally, we can ask what can be said about the particular scribes in Šaduppûm that wrote the tablets we studied in Chap. 4. It is the purpose of the following section to deal with such questions.

5.2 Whys and Hows of Old Babylonian and "Šaduppûmian" Mathematics

According to the surviving evidence, a great deal of the Old Babylonian mathematical tablets was written in educational contexts. Much, but certainly not all, of the Old Babylonian school activity was held in private residences, where a small number of young people were guided by the teacher through the study of the cuneiform writing technique, followed by language, mathematics and music. Scribal activity

[11] These are, to a great extent, the lessons of the contemporary historiography of Mesopotamian mathematics (Høyrup 1996; Robson 2009, 199).

consistent with the carrying out of educational processes has been identified in Nippur, Uruk, Ur, Kiš, Sippar, Šaduppûm, Larsa, Mari and Telloh (Sjöberg 1976). [12] Babylon and Susa might also be added (Proust 2007, 53).[13] The most important ones for the study of the mathematical curriculum of the Old Babylonian period seem to be the schools of Nippur, due to the large quantity of tablets they yielded to research (Proust 2007; Robson 2009). Thus, in general terms, it is not wrong to say that Old Babylonian scribes, either in Šaduppûm or elsewhere, wrote mathematical tablets because mathematics had to be taught to the new generations. Yet, all this will probably be nuanced by further research. For instance, the mathematics we found in the Tell Harmal tablets examined here is hardly the one found in the Nippur schools. Besides, there are many questions still unanswered: while teaching mathematics to the younger, what did Old Babylonian scribes think about it? How were they able to mediate the teaching and the applications of mathematics? And what other kinds of ideological support did mathematics have besides counting and accounting?

The previous chapters, especially Chap. 4, have made clear that a lot is known about Old Babylonian mathematics. This shows that we can learn a great deal about its vocabulary, its techniques and its favourite themes, by reading the mathematical tablets themselves. However, these tablets do not inform us about what Old Babylonians believed to be the nature of mathematical thought or the reasons why mathematics worked. The question can be addressed if we examine the so-called eduba texts. These texts attest to the importance of being able to do the duties of counting and accounting, or the duties of numeracy, but do not explicitly tell us what mathematics was thought to be.[14] Maybe this is evidence of the fact that for Old Babylonians it was much more important to be able to solve problems and deal with numerical situations efficiently than to ask what all these things were. More importantly, this may indicate that mathematics was not felt as something that needed to be explained at all: the concrete vocabulary might have made appeal to everyday life; numbers and measures in mathematical tablets might have been

---

[12] The Sippar case is commented by Sjöberg mainly in relation to the existence of female scribes and a connection to Nisaba, the patron goddess of scribal activity; no explicit detail about a scribal school is given. Tinney (1998, 49) reports that the site of Tell ed-Der, corresponding to the sister city Sippar-Amnanum, probably held a place of scribal education, the house of the high lamentation-priest Ur-Utu.

Sjöberg's reason for including Šaduppûm in this list is not directly related to the mathematical tablets. I will have more to say about the issue in Sect. 5.4.

In Sjöberg's original list, two rooms of the palace of Mari were also included, but already with a note of caution, because, among other things, until then no lexical text had been found there, or in any other dependencies of the palace. One of the rooms was eventually considered to be a storage room and not a school room (Isma'el and Robson 2010, 154).

The cities mentioned in the text are not an exhaustive list.

[13] Haddad 104 and the fragments of other mathematical tablets found in Tell Haddad and Tell es-Seeb, including school exercise texts, point to the existence of a school in Me-Turan too (al-Rawi and Roaf 1984; Cavigneaux 1999), but caution is necessary before stating this for sure (Isma'el and Robson 2010, 154).

[14] Dialogue 1, Enkimansum and Girini-isag, and Examination Text A, quoted in the previous section, are examples supporting this statement.

ultimately comparable to numbers and measures in the concrete, palpable sense; and the commitment to memory of a large amount of mathematical information during the educational process—arithmetical tables, metrological lists and tables—might have given mathematics a degree of certitude as strong as that of the language students learned through the lexical lists. In the same way that there was no need explaining everyday life, concrete numbers and measure, or language, no need was felt to establish explanations for the nature and functionality of mathematics. In some tablets, we see numerical checking of the results and in a very few some effort towards the understanding of concepts (IM52301, with its two problems and the generalisation of a procedure to calculate the area of quadrilaterals, seems to be an example of this), but never a systematic effort to produce proofs in the meaning we are used to.

A second point I am concerned about, while trying to approach the question of why Old Babylonians did mathematics, is the mediation between mathematical training and actual applications of mathematics. As just mentioned, we know a lot about the mathematics of Old Babylonian times. In the same way, thanks to the large quantities of extant contracts and administrative documents involving at least numeracy (one of the aspects of mathematics), we are allowed to say that we know a lot about some of the practical uses of mathematics. However, and this is what I want to draw our attention to, the link between mathematical tablets, administrative or other numerically oriented texts, and the eduba texts is partly the result of our inference. Although the accounting mentioned in eduba texts (as in Sect. 5.1) may have some relation with the accounting of contracts and administrative documents, it is not necessary that there was only one unified accounting practice. On the contrary, the reading of administrative documents shows many variations in the accounting practices themselves as well as in the metrological systems in use. Secondarily, the difficulty in establishing the nature of the link between the mathematics from the edubas and the mathematics as used in administration resides in that the field does not have answers to such basic things as the following: was this relation always consciously maintained by the apprentices? Is it true that every writer of an administrative text revealing a skill in numeracy acquired this skill in an eduba, maintained maybe in a private residence? Finally, what interactions among the schools themselves or between the schools and the administrative institutions could account for the degree of homogeneity (certainly, not absolute homogeneity) we find in the extant mathematical tablets?

Finally, there is an interesting relation between mathematics and the exercise of justice that served as an ideological support for the former (Robson 2008, 115–123). Understanding this relation, that I summarise in the next few lines, helps explain some of the cultural basis for mathematics, dating back to the Sumerian period and maybe having some secondary roots also in the Old Babylonian period. Eleanor Robson examined the texts from Sumerian literature present in the environment of scribal education and showed that they sustained that the exercise of justice—for instance, "the fair mensuration of land amongst the people" (2008, 122)—was to a considerable extent thought to be derived from the ability to count and measure. A second part of the argument starts by analysing two similar types of images: firstly, of gods giving kings a pair of symbols related to kingship and justice, the rod and the ring (as in the head of Hammurabi's stela); secondly, of naked goddesses holding these symbols. In a special case, the Burney Relief, also called the Queen of the Night (which, by the way, illustrates the cover of Robson's book), the rod and the ring have been interpreted as a "1-rod reed and a coiled-up measuring rope" (Robson 2008, 120), suggesting thus a fusion of two different but parallel motives. The argument is finally completed with the consideration that in royal hymns "Nisaba bestows the reed and the rope on kings as symbols of literate and numerate justice" (Robson 2008, 120). Thus, merging all these ideas, it is a possible conclusion that knowledge of mathematics was believed to be ultimately provided by the gods, and it was in their interest that this knowledge must be used, especially by kings, in the distribution of justice.

To sum up, the question of why Old Babylonian scribes developed and maintained a tradition of mathematical studies leads us to considerations about practical necessities as well as transcendental matters. On the one hand, the ability to count and to measure was endorsed by its utilitarian character, as can be attested in both the eduba texts and the many contracts and administrative documents from the period. On the other hand, this utilitarian character itself could be praised because of its service to the exercise of justice and kingship. None of these factors, however, can explain the specific details of how the system of schools gained stability, how it interacted with the activities of working scribes and how the specific form of the Old Babylonian mathematical discourse came to be. The mentioned factors help us understand, rather, the structural relations that supported mathematics in Old Babylonian culture and not its implementation details.

With relation to the last point, we may also notice that there is no warrant that Old Babylonian mathematics, as a body of knowledge, was the result of the sole efforts of the edubas. A full explanation for the shaping of that mathematical tradition should contemplate the knowledge about numbers, measures and shapes that might have been brought to the edubas from external groups as the surveyors (see Høyrup 2002, 378ff) and from oral traditions in general (see again Høyrup 2002, 362ff).

Even if we leave aside the question of the shaping of Old Babylonian mathematics, still there is the problem of its permanence that must be accounted for by historiography. Apart from the usefulness of mathematics and its role as a justice device, it would be interesting to know the less structural factors that kept it alive.

In relation to the 12 tablets studied in this work, there are a few things that can be said. Šaduppûm was an important administrative center of the kingdom of Ešnunna. This is shown by the large number of texts of administrative character exhumed from the site, as well as by the prominence of the main administrative building in the city (Baqir 1959; Hussein 2009). Furthermore, some cultural importance of pre-Hammurabi Ešnunna is highlighted by the fact that Mari, when reforming its writing system, emulated Ešnunna, not some centre in the south (Charpin and Ziegler 2003). Ten (or maybe nine; see Table 1.1) of the mathematical tablets examined here come from the same room 252 of the site. In principle, if we consider that Šaduppûm does not break with the general patterns of Old

Babylonian mathematics, that is to say, if we assume that the mathematical tablets from Šaduppûm are connected to a place of scribal education, then we are entitled to ask whether there was an eduba at the city. Unfortunately, without access to the yet unpublished excavation reports, a secure answer seems out of the reach. This does not prevent a little guess work though, not for the sake of mere speculation, but in order to help set productive questions for future research, mine or others'. Part of the fund of more than 3000 tablets exhumed from Tell Harmal has not been published yet. Hussein (2009, 92) estimates that around 600 texts have come to light. Apart from the 12 mathematical tablets studied here and previously published by Taha Baqir in *Sumer*, Isma'el and Robson (2010) report that Tell Harmal produced only four other known tablets of the genre. Two constitute a pair of variants of a mathematical compendium, the other two are arithmetical tables. To this, we should add the typical school tablets that al-Fouadi (1979) published, of which a provisional list would include: passages from metrological lists (IM51750, IM54987-A, IM63106, IM63143 and IM67330), a list of reciprocals (IM54987-B) and many lexical texts. Among the 2000 tablets not yet published, more school tablets may be found and maybe more problem texts. However, even with the present attested evidence, small as it is, it seems to me that this group of texts points to the possibility that Šaduppûm hosted a place of scribal education.

5.3 The Genetics of the Mathematical Practices: How Did Old Babylonian Mathematics Evolve?

One frequently used approach to the study of the development of Mesopotamian mathematics is the analysis of formal characteristics of mathematical products, in the domains both of the discursive practices, including language traits, and the materiality. The main assumption behind this procedure is that certain characteristics of mathematical products are hereditary—in other words, when one writes a mathematical tablet, one tends to reproduce certain language uses, disposition of the words on the surface of the tablet, shape and size of the tablet, themes for word problems and technical procedures that were learned previously by means of other mathematical pieces. A second, maybe equally important assumption in this line of reasoning, is that when mathematical tablets from distinct localities show repetitive or related similar characteristics, it is possible to assume some form of kinship between the contexts that produced them: either an influence of one on the other or a common ancestry. In other words, there is a genetics of mathematical practices.

As we know, cut-and-paste geometry, scaling of figures, sexagesimal place value notation and reciprocals are examples of characteristics of a large range of Old Babylonian mathematical tablets. The development of the sexagesimal place value notation occurred in the second half of the third millennium, in a very non-linear, complex process involving strictly mathematical as well as administrative practices. The second half of the third millennium is believed to be a key period in this development. Although there are few mathematical tablets from the period, the steps by which the sexagesimal system took its form can be traced in the texts of other genres, especially the economic and administrative.[15] In the Old Babylonian period, sexagesimal notation had become ubiquitous. Reciprocal tables had already been present in the Ur III period. All but one (IM54559) of the tablets analysed in Chap. 4 use reciprocals. The scaling of figures and the cut-and-paste geometric procedure, on the other side, are present with less frequency, but they are so recurring and so well distributed through the geography of ancient Mesopotamia that they too can be taken as general characteristics of Old Babylonian mathematics.

Thus, these aspects show the affiliation of the tablets from Šaduppûm to the ensemble of the Old Babylonian mathematics: they share important hereditary features with the remaining of the Old Babylonian mathematics. However, this statement does not specify the role Šaduppûm played in its interaction with other polities where mathematics was also practiced. How did the scribes of Šaduppûm learn their mathematics? Did they influence scribes of other places, such as Mari, that adhered to southern scribal practices through Ešnunna (Robson 2008, 127; Charpin and Ziegler 2003)? Although appropriate answers to these questions are out of the reach at the moment, due to the lack of documentation, some interesting points can be made by analysing characteristics of these tablets that cannot be considered common to the whole of Old Babylonian mathematics, as in the following paragraphs.

As we saw in Chap. 4, a group of ten tablets has as its opening sentence the expression "If someone asks you thus, saying this"—with the possible omission of the pronoun "you". This is not a common feature of Old Babylonian mathematical texts in general, and it must be noticed that a very similar opening sentence is present in tablet Db_2-146, from the site of Tell edh-Dhiba'i, near Tell Harmal (Baqir 1962, 11ff). This, together with other characteristics, has permitted the field to speak of a group of mathematical tablets representing a branch of the Old Babylonian mathematics in the Diyala region.[16] But even here, a distinction can be made: IM55357 and IM52301 do not use that opening sentence, suggesting—with other characteristics—that the mathematics of that region had its own sub-varieties. More about this in Sect. 5.5.

In the domain of materiality, things seem to be a bit more uncertain. The most clear example of how material features of mathematical tablets are not arbitrary comes from the already mentioned schools of Nippur. As is well known today by the field, the shapes of the tablets produced during the education of a scribe—specifically those containing either mathematics or lexical lists or literature—are related to their pedagogical functions.[17] Put in a simple way, there are tablets for the students

[15] For the history of the sexagesimal place value notation, see (at least) Powell (1976), Friberg (2005) and Robson (2008, 75ff).

[16] See Høyrup's detailed and precise descriptions of the Ešnunnan texts (his group 7), divided in Groups A and B (2002, 319ff). See also Isma'el and Robson (2010) and Gonçalves (forthcoming) for a further characterisation of the mathematics of the Diyala.

[17] See Civil (1995) for tablet types in the domain of the lexical lists. See Proust (2007) and Robson (2008) as regards mathematics, specifically at Nippur. Tinney (1998) is an illustrated gentle exposition of the theme.

to exercise parts of a text, for instance a metrological list, tablets for the students to exercise a complete text, tablets where a model made by the teacher must be copied by the student and tablets with arithmetical exercises. These tablets come in different shapes, and even the disposition of the text is different, according to the function they have in the above list.

In Šaduppûm, as we have seen, IM55357 and IM52301 are written in the portrait format. Both have only one column. IM55357 is written only on the obverse, but the text is not really complete, so that it could have continued on the reverse. In IM52301, the left margin is used too. This is compatible with the evidence from Old Babylonian mathematics. In Robson (2008, 107), a table summarising the types of tablets in the advanced curriculum explains: "type S" is a "small, single-column tablet in portrait orientation" and its typical contents may be a "single word problem with worked solution".

The other ten tablets studied in this book, however, were written in landscape orientation. They too contain only one column and are written both in the observe and the reverse. Their contents, again, is no doubt advanced mathematics. In the summary by Robson (2008, 107), we read: "type L" is a "small, single-column tablet in landscape orientation" and typically it contains a "table of powers".[18] This is not the case of our ten tablets. Thus, in Šaduppûm at least, advanced mathematics in the form of word problems could be present both in the portrait and the landscape format. In any case, as Isma'el and Robson (2010, 162 n20) remind us, both in MCT and in TMS there are examples of word problems written in the landscape format.

To end this section, I would like to relate the two material variants with the two linguistic variants of the Ešnunnan group of mathematical tablets described by Høyrup (2002, 319ff). On the one hand, the ten landscape tablets contain advanced mathematics and represent one linguistic variant of the Ešnunnan group; on the other hand, IM55357 and IM52301 contain advanced mathematics too, but are written in portrait orientation and represent a different linguistic subbranch. As a result, there seems to be a correlation between linguistic form and material form, desirably to be confirmed by future research, if more tablets come to be known. This correlation must be taken into account when trying to link shape and pedagogical function.

The above paragraphs show how the study of formal characteristics can help gain insight about Old Babylonian mathematical tablets, from the point of view of both text and materiality. The procedure is not entirely different from the one textual criticism uses to establish an edition of a text, when a number of variant manuscripts coming from a hypothetical original is available: one assumes that manuscripts are copied from manuscripts and that the variations they show tend to be passed on from one generation of manuscripts to the next; a genealogical tree can be set up; the common ancestry is deduced from the features of the descending generations and a history of the modifications on the original text can then be written. The procedure is also present in the work of archaeologists, when artefacts are grouped into categories based on their traits. Starting from this, it is possible to characterise

[18] To this, Robson also adds the famous Plimpton 322, a small tablet in landscape format, containing a numerical table.

a material culture and to establish lineages of transformations and, thus, relative chronologies. Again, the fundamental assumption is that traits, be they of pottery, basketry or any other type of artefact, can be inherited. As a final remark, the use of typological, genetic arguments in the study of the mathematics of Mesopotamia, textual criticism and archaeology points to common ancestry or shared interests or even peer interaction between the disciplines. This, of course, is not to be taken as sheer coincidence, but it is an evidence of the way they interacted while establishing themselves as fields of knowledge.

5.4 The Mediation between Mathematical Artefacts and the Larger Society

An additional approach to the study of mathematical texts and tablets inquires for the relations between mathematics and the larger society. It is needless to say that this approach assumes that there is a two-way road that permits mutual influences between mathematical practices and more general social, cultural and economic issues. One should also observe that mathematical practices are taken here under two different points of view: at the same time that it is known that they are inseparable parts of the society in which they are imbedded, they are occasionally seen as if detached from the society and thus prone to influences external to them. The following paragraphs bring a contribution to the problem in relation to both Old Babylonian mathematics in general and the tablets from Tell Harmal in particular.

An example that is commonly adduced to treat this problem is the presence of common life themes in mathematical tablets. As we have seen in Chap. 4, brick wall construction, exchange price of lard and oil, assignment of a task to workers, measurement of a field and the determination of the capacity of a vessel are, so to say, daily matters that find their presence in mathematics. One may object that they are not exactly utilitarian issues, on the grounds that numbers in mathematical tablets are too perfect to reflect actual everyday situations and, more important, that the involved mathematical contents do not occur outside mathematics.[19] In order to illustrate the question and gain insight into the ways mathematics accommodates empirical issues, I will quote from the Laws of Ešnunna and the Letters from Harmal. But first some observations are required.

The Laws of Ešnunna became known through a pair of tablets exhumed during the excavations conducted by the Directorate-General of Antiquities of Iraq in the 1940s. This pair of tablets comprises two variant copies of the same set of laws and was discovered in two different rooms at stratigraphic Level II, being thus, in general terms, of the same period as most of the mathematical tablets studied here (only IM55357 came from Level III; all others were found in Level II).

[19] Related to this issue, Høyrup (2002, 384) introduced the term *supra-utilitarian* that refers to situations with artificial structure and not to be expected to turn up in practice (for instance, to know the area of a rectangular field but not both sides).

One of the copies of the laws (Text B, tablet IM52614) is dated according to stratigraphy to the reign of Daduša (Goetze 1948, 66). The other copy (Text A, IM51059) explicitly mentions the name of Bilalama, a king of Ešnunna that "lived only a short time after the downfall of the Third Dynasty of Ur" (Goetze 1948, 66–67). Due to the variant readings and the orthographic errors these two tablets bring, it may be inferred that they were not official copies, but rather "the product of a school of scribes" (Bouzon 2001, 15).

The Letters from Harmal are a set of 50 letters found in the site of Tell Harmal. Most of them came from stratigraphic Level II and only two from Level III (Hussein 2009, 94–96). According to Goetze, the stratigraphic dating can be confirmed by a mention that Letter 5, a royal letter, makes in its envelope to Ibal-pi'el II, son of Daduša and the last king of Ešnunna (Goetze 1958, 5).

In this way, our mathematical tablets, the set of laws and the letters came all from the same archaeological site and from the same stratigraphic levels—with a concentration at Level II. As a consequence, it is possible to assume that the contexts, groups and generations of scribes that produced them had much in common and may have at least partially coincided.

In the opening section of the Laws of Ešnunna—that I freely quote from Goetze (1948) and Bouzon (2001)—we are informed about the prices of many commodities, probably the highest price a dealer could charge for them (Bouzon 2001, 61), including lard and oil: "1 ban_2 2 $sila_3$ of oil for 1 shekel of silver", "1 ban_2 5 $sila_3$ of lard for 1 shekel of silver". Comparing to what we find in IM54464 (see Chap. 4), we notice that the price of lard is the same in the laws and in the mathematical tablet; the price of oil is a little lower in the laws (in IM54464, 1 shekel of silver buys only 1 ban_2 of oil). This all shows that in IM54464 the scribe not only took oil and lard from a set of usual commodities but also fixed their imaginary prices for the mathematical problem in the vicinities of the prices that could be practiced in daily life. The problem gained, in this way, a strong sense of reality. The two-way road mentioned in the opening paragraph of this section enabled scribes to transpose to mathematics a portion of the larger reality; at the same time, as they themselves were part of this reality, they produced an imaginary situation by which their young apprentices could develop the skills necessary to conduct life as it was in the enlarged context.

As a second example, the so far unreadable IM54010 (see Chap. 4) seems to deal with harvest issues. A form of the verb *eṣēdum*, to harvest, apparently occurs twice in it. This theme is also present in texts outside mathematics. To quote again the Laws of Ešnunna (Goetze 1948, first part of §9; Bouzon 2001, §9), we read: "if a man gives 1 shekel of silver to a hired man for harvesting and if he does not fulfill his obligation and does not complete for him the harvest everywhere (or during the whole harvest), he shall pay 10 shekels of silver". The theme of harvest is also present in the Letters from Harmal (Goetze 1958). In Letter 31, we read: "Concerning the socage due to the palace which you are accustomed to perform yearly at harvest time, have my tablet read and come here quickly with four oxen of yours. Should you not come you will measure out 1 kur of barley per bur of field which falls fallow". Thus, it can be suggested that harvesting was culturally associated with numbers and measures not only in mathematical texts. Admittedly, the presence of numbers and measures in the Laws of Ešnunna—and for the same effect in the Laws of Hammurabi—is well known, and it is not exclusively related to harvest issues. However, what I want to emphasise here is the use that the scribe of IM54010 made of culturally available material, the association between harvest and calculations.[20] It is also striking that the expression "1 kur (*kurrum*) of barley per bur (*burum*) of field" in Letter 31 is reminiscent of "in a *burum*, how much did I harvest" and "the *bu-ur* of your field" that appear in the mathematical tablet. Once more the two-way road was in operation.

In Old Babylonian mathematics, the presence of daily life is quite strong and long known and well attested. Nemet-Nejat (1993, 149–249) lists hundreds of terms and combinations of terms in a "Glossary of Realia from Cuneiform Mathematical Texts". It remains for the field to investigate the specific transit each term, each concept and each idea had while circulating between the larger reality and the realm of mathematical tablets. Such an investigation will perhaps reveal differences according to the periods and places we select. The previous paragraphs, focusing on the tablets from Harmal, are a contribution to the question.

From the point of view of materiality, there is still much to be understood, with the result that apparently only general statements can be made at the moment. The schools of scribes, the edubas, as already commented, were not large buildings made for the purpose of hosting the educational activities. The learning necessary for a young person to become a professional scribe was perhaps given mostly in private homes. The school that functioned in House F of the site of Nippur had material evidence of educational activities: a large number of students tablets, bins for recycling clay, raw clay for making new tablets, an open court for the lesson and a few benches for the students (Robson 2009, 202). However, it is not certain that every school of scribes had to be so conspicuously equipped. To my knowledge, no report of such facilities in Šaduppûm has come to press yet. Nevertheless, the assumption that some kind of schooling existed in that city can be backed firstly by the existence of both elementary and advanced mathematical tablets, as well as lexical texts, as mentioned in Sect. 5.2. A second reason is given by Sjöberg (1976, 177): "The finding of lexical texts and literary texts and geographical lists shows scribal activities in this city. It should be noticed that the texts were found in the temple dedicated to Nisaba and her spouse Haja, patrons of scribes and scribal art".

In the same way, research still does not know much about the circulation of mathematical tablets. What were the mechanisms that could make possible for a tablet to be taken elsewhere? Is it true that the group of nine or ten tablets found in room 252 was produced in that room? These are open questions at the moment.

Finally, recycling and discharging are also a matter of interest. In the site of Nippur, always the best informed case, a great portion of the tablets from House F was found as part of the walls, the floor and the furniture of the house, showing their use as building material after their pedagogical function had been completed

[20] An association that is also present in the harvest documents of Tell edh-Dhiba'i, in the vicinities of Tell Harmal (Sulaiman 1978).

(Proust 2007; Robson 2009, 202). Also in House F, there were pots that functioned as recycling devices for tablets that, once written, had no other use. Whether a tablet was used as fill or recycled, on the one side, or conserved, on the other side, depended possibly on its contents and expresses how this contents were valued.

A last approach that is concerned with the place of mathematics in the larger society starts by considering it as part of social and economic history. In this way, the initial force that motivated the production and development of what would be afterwards Old Babylonian mathematical knowledge was the necessity of administration that rose with the advent of the first cities and states, during the fourth millennium. This process was closely connected with the development of writing, and the field has already passionately discussed which phenomena drove the others.[21] In any case, there seems to be a consensus that the needs of administration and bureaucracy were a proper environment for mathematical practices to develop, specifically in order to do counting and accounting, as mentioned in the eduba texts.[22] Temples, and afterwards palaces, had to keep track of the quantities, for instance, of grain, beer and oil they had in their deposits; the quantities given to workers as rations or payment for services and the quantities received from producers. The mathematically capable administrator was the person to do the task (Høyrup 2002, 311–316; Mieroop 2007, 26; Robson 2008, 38–40). There are also indications that there existed mathematical practices not necessarily linked to temple, palace and state administration. Whether bureaucracy and administration were the only soil where mathematics germinated is difficult to know. It is possible that merchants' and lay surveyors' varieties of mathematics existed (Høyrup 2002, 362ff), but the evidence for this is very indirect.[23] Anyway, the relationship between mathematics and state (and, in this context, justice too) should not be taken as a fixed one. If before the Old Babylonian period mathematics developed in close dependency of and influence on practical matters of state administration, the Old Babylonian period saw the emergence of a more independent type of mathematics, in a specific sense, humanistic (Høyrup 2009).

---

[21] See Høyrup (2002, 311) for an abstract of the question and further references. See also Høyrup (2009) for an enlarged development of the theme.

[22] If this is correct, then the eduba texts, as used in Old Babylonian times, may refer mostly to the initial stages of mathematical practices. As a consequence, trying to match a description of Old Babylonian mathematics with the mathematical practices mentioned in these texts is a misguided goal. It is to be noticed the following parallel: speaking about the Old Babylonian schools of scribes, George (2001, 7) states that "to attempt to identify the many private houses where scribes were trained in the Old Babylonian period with the grand institutions called é.dub.ba.a in Sumerian literary texts is misconceived. To look for material remains of this é.dub.ba.a in Old Babylonian levels is to try to match realities of two different eras".

[23] In a very careful and involved reasoning, Høyrup (2002, 362ff) sustains that the later Indian, Greek and Islamic mathematical traditions must have borrowed a certain set of characteristics they had in common with Old Babylonian mathematics from the same "source that also inspired the Old Babylonian school" (2002, 374). After arguing that the Ur III school could not be this source, he concludes that the only alternative left must have been a lay tradition. The evidence is, in this way, indirect and it is based on the existence of common ancestry for different mathematical traditions with a certain degree of regular, repetitive features.

5.5 Final Remarks and Trends for Research

The concern of the field to understand the geographical variations that comprehend Old Babylonian mathematics is not new. In 1945, Goetze published his study "The Akkadian Dialects of the Old Babylonian Mathematical Texts", as a chapter of Neugebauer and Sachs's MCT. In it, Goetze separated Old Babylonian mathematical tablets into different groups defined mostly by their linguistic traits, especially the way each group employs the cuneiform signs in spelling. According to that study, there would be six different strands of mathematical language usage, representing four varieties of Akkadian: a southern variety from Larsa, another southern variety probably from Uruk, a northern variety, and one with northern modernisations of southern texts (MCT, 151). A more detailed description, taking into account newer texts as well as some that Goetze did not want to consider (texts with heavy logographic use), was presented by Høyrup (2002, 319ff). The new study combined the orthographic criteria used by Goetze with "observations of terminology in the widest sense (vocabulary in the context of function)" (Høyrup 2002, 319). The results include modifications on Goetze's classification and a new group, the Ešnunnan texts, divided in an homogeneous subgroup 7A (our ten landscape tablets belong here) and a more diversified subgroup 7B (IM55357 and IM52301, our two portrait tablets, belong here, with other four tablets from Tell Harmal—two variants of the mathematical compendium, IM52916 and IM52685+52304, and two arithmetical tables, IM43993 and IM121613—, the quoted Db_2-146, from Tell edh-Dhibaʾi, and Haddad 104, from Me-Turan). Ismaʾel and Robson (2010) published a few texts from the Diyala region, specifically from Tell edh-Dhibaʾi and Tell es-Seeb (plus an unprovenanced one). The authors offered a general, preliminary characterisation of the mathematics of the region, as a "contribution to a historical geography of Old Babylonian mathematics that is now emerging" (2010, 160). In Ešnunna there would have been a unique break with the typical Old Babylonian mathematical genres as known from other places: this is exemplified by IM52301, which contains word problems and a coefficient list. Furthermore, the authors' state, the orthographic habits of the mathematical tablets from the region and their terminology may indicate an existence of this branch of Old Babylonian mathematics that would have been independent from the southern tradition. In this context, I would like to add that a full geography of Old Babylonian mathematics would profitably include samples of the locally culturally available material that transited to mathematics, in order to strive for better understanding of the mediation processes that were in effect between the semantic composition of mathematical texts and the communities where they were produced. My mentions of the Laws of Ešnunna and the Letters from Harmal, in Sect. 5.4, are a contribution to the research.

A complementary aspect of a geography of mathematics is the search for the factors that made it possible for Old Babylonian mathematics to spread over so vast a region. More than often, the dispersion of mathematics over significant parts of Mesopotamia is taken as a given, almost as a consequence of the dispersion of cuneiform and other cultural institutions. Although this assumption is true, it avoids the question of explaining the presence of mathematics in each case, or at least how the inception of new vocabulary or techniques occurred in different places. Certainly, the communities of scribes were engaged in some sort of exchange. We know that letters circulated—after all that is what letters are made for—and it is also clear that mathematical knowledge circulated, once it is found in many different places, but it is not clear how this was performed: were mathematical tablets carried over from one city to another? Or was mathematical wisdom something to be carried on one's memory? This leads us to the issue of the adequate methodologies to deal with the question. It is possible that the mathematical interactions among Mesopotamian cities can be understood only in the large scale—for instance, as a possible application of a theoretic framework like the peer polity interaction model (Renfrew 1986; Cherry 2005). In this case, research could understand better the processes of the spread of mathematical knowledge as a consequence of the development of other, leading institutions such as the monumental building and the exercise of justice, where competition and competitive emulation—requirements for application of the model—among the interacting members can be more easily located. We would, then, know more of the spread of mathematical techniques and we would be able to assess from a new point of view the place it was given in Old Babylonian culture.

Beyond mathematics, mathematical tablets may also contribute to the study of the devices that Old Babylonians developed to assist in their cognitive processes, as part of a *cognitive history* of Old Babylonia, much in the same spirit of Reviel Netz's cognitive history in the context of the shaping of deduction in ancient Greece (Netz 1999). Mathematical tablets, together with lexical lists and the Sumerian compositions that constitute the curriculum of the schools of scribes, inform us about the nature of the cognitive operations Old Babylonians made in order to understand and modify the world around them. The classificatory character of lexical lists is akin to the list of coefficients—of which IM52301 brings only a small sample. Cut-and-paste geometry and scaling of figures are also tools for the cognition that cannot be separated from their concrete, almost palpable way of dealing with lines and areas in order to solve mathematical problems. The compromise of Old Babylonian mathematics with concreteness might be seen there too. Literature in the field has already dealt, although mostly in an indirect way, with many of these problems. To mention only one example, Niek Veldhuis's (1997) approach to lexical lists considers, as secondary elements of its main reasoning, a number of issues related to cognition. A study of cognitive tools specifically in the context of Old Babylonian schools would encompass mathematical as well as lexical and Sumerian texts. In this respect, besides all the geometrical apparatus, as in cut-and-paste geometry and the scalings of figures, certain language usages, for instance, some specific syntactical patterns that appear in mathematics, divinatory and medical texts (Ritter 2005), might offer a cognitive background for mathematics.

Mathematical tablets are part of the broader spectrum of *history of science*, a somewhat anachronistic and positivist name for a discipline that could be better named *history of knowledge*. They are also pieces of the larger cultural, social and economic history of Mesopotamia. One of the main difficulties in integrating mathematical artefacts in those larger contexts is that with astounding frequency they have been taken as *the* starting point of investigations. As a textual corpus they are given, the researcher limits his or her questions to what this corpus contains. However, history can also profit from an inversion of these elements. A question may be the starting point of a research and feed the constitution of a textual corpus to answer it. This corpus may contain mathematical texts or not, and it will be much better if it is a hybrid set of objects, as mathematical texts together with laws and letters (as exemplified in Sect. 5.4) or mathematical texts together with lexical lists and Sumerian compositions (as in the studies by Proust and Robson about the elementary curriculum of Nippur). I am not advocating here a blind, strict adherence to an *histoire-problème*—even its practitioners cannot deny that the study and contact with a previously established textual corpus are necessary to formulate the *problème*, as Duby (1991) clearly made us aware in his intellectual autobiography. Evidently, there will always be the necessity of textual studies, as a step to build the possibility of reading the texts. It is also evident that the cuneiform mathematical corpus can pose and has posed interesting questions, such as the role of geometrical thinking and the distinction between the abstract and concrete uses of numbers. Yet, these texts came to life in integration with other cultural and material productions of Old Babylonia, so it is natural to devise historiographical questions that are not necessarily answerable only through the mathematical evidence.

Chapter 6
General Vocabulary

The following vocabulary lists all words that appear in the corpus. For each of them, I present a semantic field in which it fits for the purposes of the present work and the translation I adopted, followed by the occurrences of the word in the analysed corpus.

The idea of dealing not only with a translation but also with the semantic field is that the chosen translation should ideally have some proximity with the nuances the word has in general Akkadian or Sumerian texts, given by what I call "semantic field". Even though it is true that some mathematical terms might be considered too technical to present such link between its uses in mathematics and in other contexts, for many of our terms the relation does exist.

One important further idea to have in mind is that besides the semantic field and the translation, we can sometimes speak of the meaning in context of a word. The difference is that the adopted translation is an English word or expression that, in principle, complies with requirements of clearness and conformability (as in conformal translations; see Sect. 3.1), whereas the meaning might be an explanation or a general idea, more freely expressed in a modern language.

An example might be useful to understand what is meant here.

In Akkadian, the verb *kamārum* is used to express the action of accumulating. Its semantic field is characterised by a range of meanings that are more or less associated: to heap up, pile up, to spread, to accumulate, to have in store (CAD, K, 112).

When using this verb in mathematical texts, scribes seemed to be coherent with the meanings belonging to the semantic field. Specifically, *kamārum* might mean the action of putting two segments or numbers side by side and considering the total amount given by them. While this may be considered an addition, it has the specific tone of being generally a symmetric one (in relation to the two added things) and tends to be used when things added are considered only in their numerical values, either having different natures or not. This is its meaning in context.

Finally, when dealing with mathematical texts that are to be translated into a modern language, one has to choose an equivalent or at least a satisfactorily equivalent term to

C. Gonçalves, *Mathematical Tablets from Tell Harmal*, Sources and Studies in the History of Mathematics and Physical Sciences,
DOI 10.1007/978-3-319-22524-1_6

express the intended meaning and its membership to the semantic field (when this is the case). Besides, one tries also to comply with the requirements of conformability and readability. In the case of the verb *kamārum*, the adopted translations is "to accumulate".

In order to make the use of the vocabulary more coherent with the textual corpus I have dealt with, some words are recorded here under headings different from the ones they have in the CAD. This is because CAD consistently uses late writings, while in this book I have dealt with Old Babylonian writings. Some examples are *awīlum* instead of *amīlum* and *wabālum* instead of *abālum*. Furthermore, in some cases, where there seems to be a writing specific to Ešnunna, this is used: for instance, *sattakkum* is preferred over *santakkum*, because the former seems to have been the writing used by our scribes.

After the semantic field, as already explained, all the occurrences of the term in our tablets are enumerated, obeying the following rules: square brackets indicating damaged signs on the tablet are maintained; question marks indicating doubtful readings are also maintained; exclamation marks are maintained too; << >> indicating parts of the text that are to be deleted in the reading of a tablet are kept here only in one case, *me-eḫ-<<ša>>-ra-am*, because the deletion corrects a wrong spelling and insertions, indicated by < >, are always kept here.

Finally, after the occurrences of the word, I write in some cases an explanatory comment. I have tried to leave in this chapter only the comments that are aimed at clarifying some more intricate points in my procedure. Commentaries dealing with problems still under discussion by the field are presented and discussed in Sect. 3.2.

For each Akkadian word except the functional ones (*ana* and *ina*, among others) and units of measure, there is in most cases one English word or short expression selected as its translation in the present work. In all cases, the adopted English translation is underlined, so that the reader can easily identify it.

Finally, there are three Sumerian entries, ib_2.si_8, ib_2.si.e and igi, for which no Akkadian equivalents were used in this work.

akālum v.; to eat, consume, take for oneself, use, enjoy

šu-ta-ki-il-ma, IM52301
tu-uš-ta-ki-lu, IM52301
šu-ta-ku-il-ma, IM52301
tu-uš-ta-ka-al-ma, IM52301
šu-ta-ki-il-ma, IM53957
šu-ta-ki-il-ma, IM53965
šu-ta-ki-il-ma, IM54559
tu-[*uš*]*-ta-ki-lu*, IM54559

Comment: The Št-Stem is translated in this work as to combine. See also the discussion in Sect. 3.2.

alākum v.; to go, come, arrive, move

al-li-ik, IM53965
[*a*]*l-li-ik*, IM53965

amārum (igi.du$_8$) v.; <u>to see</u>, examine, experience, find, come to know

igi.du$_3$, IM55357
ta-mar, IM52301
ta-m[*ar*], IM52301
[*ta*]-*mar*, IM52301
ta-mu-ru, IM52301
[*t*]*a-mar*, IM52301
ta-mar, IM54478
ta-mar, IM54538
ta-mu-ru, IM54538
ta-mar, IM53957
ta-mar, IM54010
ta-mar, IM53965
ta-[*mar*], IM54559
ta-mar, IM54559
ta-mar, IM54011
ta-[*mar*], IM54011

ammatum (kuš$_3$) n.; forearm, <u>cubit</u>

am-ma-tim, IM53961
am-ma-at, IM53961
kuš$_3$-*ia*, IM54010
am-ma-at, IM53965
am-ma-tim, IM54011
kuš$_3$, IM54011
[*am-ma-tim*], IM54011

ana (nam) prep.; to, for, up to, towards, against, upon

a-na, IM55357
nam, IM55357
[*a*]-*na*, IM52301
a-na, IM52301
[*a-na*], IM54478
a-na, IM54478
a-na, IM53953
[*a*]-*na*, IM54538
a-na, IM54538
a-na, IM53961
a-na, IM53957
a-[*na*], IM53957
a-na, IM54010
a-na, IM53965
[*a*]-*na*, IM53965
a-na, IM54559

a-[na], IM54464
a-na, IM54464
a-na, IM54011

Comment: except for an incomplete sentence (IM54559, line 2), the preposition ana is used only once with the translation towards (IM54011, line 3) and, in all remaining occurrences, it appears with one of the verbs ***našûm*** and ***waṣābum*** and it is translated as to.

aramanītum n.; a mathematical term

a-ra-ma-ni-a-ti-a, IM52301

Comment: kept untranslated *aramanitum*.

arkum (gid_2) adj.; long, tall

gid_2, IM55357

ašlum n.; rope, tow rope, surveyor's measuring line

a-ša-al, IM54538
a-ša-al, IM54011
a-ša-[al], IM54011

Comment: used with the meaning of a measuring unit

atta (za.e) pron.; you (masc. sing)

za.e, IM55357
za.e, IM52301
at-ta, IM52301
at-ta, IM54478
[*at-ta*], IM53953
at-ta, IM54538
at-ta, IM53961
at-ta, IM53957
at-ta, IM54010
at-ta, IM53965
at-ta, IM54559
at-ta, IM54464
at-ta, IM54011

awīlum n.; human being, man, person, somebody, grown man, male, free man

a-wi-lu-ka, IM54538
a-wi-li-im, IM53961
a-wi-lim, IM54011

bamtum n.; half

ba, IM55357
ba-a, IM54010

Comment: See the commentary to IM55357 and Høyrup (2002, 31, note 53) for a discussion of the term and the writing *ba*.

banûm v.; to build, construct, form (a city, building, wall), make, shape (a stela, statue)

ab-ni, IM52301

basûm (*basi-*) n.; the equalside, square root, cube root, side, edge, the equal

ba-se-e, IM52301
ba-si-ka, IM52301
ba-su-šu, IM52301

Comment: See Sect. 3.2 for a discussion of this term.

būdum n.; shoulder, part of the body between the shoulders

murgu$_2$, IM55357 (see also *warkûm*)

Comment: the CAD (B, 303, s.v. *būdu*) attests murgu as a logogram for *būdum*. The sign in IM55357 is most likely murgu$_2$ (see also Høyrup 2002, 231). According to the ePSD, both murgu and murgu$_2$ are used as logograms for *būdum*, written however as *pūdu* in the ePSD (s.v. murgu). According to the CAD, *pūdu* is a variant of a second *būdu* or *bu'du*, all with more or less uncertain meanings, but apparently not related to "shoulder". However, the CAD (E, 344, s.v. *eṣenṣēru*) also reports the following entry in lexical list A=*nâqu* (V/1 84ff): mur-gu SIG$_4$=*pu-u$_2$-du ša$_2$* [*amēli*], shoulder (but this is a late lexical list). Interestingly, the same passage is quoted as *bu-u$_2$-du* in another place (CAD B, 303–304, s.v. *būdu*). An Old Babylonian testimony is brought by Proto-Izi 275ff (quoted also in CAD B, 303–304, s.v. *būdu*). All in all, it seems that we can assume, in the present context, that both *būdu* and *pūdu* are acceptable spellings for the word, the ambiguity being due above all to the possible phonetic values *bu* and *pu* of the initial sign in the syllabic writing. Finally, Borger does not include any logogram corresponding to the entry *būdu* in the Glossar of MesZL; instead he refers the reader to the discussion of signs murgu$_2$ (901) and murgu (906) and suggests that the correspondence between these signs and the words *būdu*, *ṣēru* and *eṣenṣēru* could be further discussed. For the present purposes, I assume that both murgu and murgu$_2$ could be used in Ešnunnan texts, given the frequency of unorthographic writings in the region.

burum n.; a surface measure

bu-ur, IM54010

elēnum adv.; above, over, upstream, besides, apart from, beyond

e-le-num, IM54011

eli prep.; on, above, over, against, more than, beyond, at the debit of, on account of, on behalf of

e-li, IM52301
e-li, IM54559

Comment: in both IM52301 and IM54449, *eli* is used with the verb ***watārum***. The expression thus formed is translated as “to go beyond”.

ellum (i_3.giš) n.; sesame oil (of a specific quality), (a good) sesame oil

i_3.giš, IM54464 (see also *šamnum*)
ul-li-k[*a*], IM54464
ul-li-im, IM54464

Comment: Neither the CAD nor the AHw registers the writing with the initial *u*. However, as the tablet also uses the logogram i_3.giš, it is reasonably safe to assume that, at least in the present context, *ullum* = *ellum*. The CAD (E, 106, s.v. *ellu* B) reports that “only the series Hh. consistently distinguishes I_3.GIŠ = *ellu* from I_3 = *šamnu*”, which reinforces the identification *ullum* = *ellum* in the Old Babylonian period. As the syllabically writing points to *ellum*, we can assume that it is this word that is at issue here and not *šamnu*.

elûm (an.ta) adj.; upper, up above

an.ta, IM55357
e-lu-um, IM52301
e-li-tum, IM52301
a-li-a-am, IM52301
a-li-am, IM52301
e-li-tim, IM52301
e-li-im, IM53953
e-li-tum, IM53953
e-lu-um, IM53953

elûm v.; to travel uphill or to higher ground, come up, move upward, rise, grow, emerge, come out, to show up, turn up, appear

šu-li-[*ma*], IM52301
šu-li-ma, IM52301
i-li-ku, IM52301
i-li, IM53953
i-li-a-ku-um, IM53953
[*i*]*-li-ku-um*, IM53953
i-li-ku-um, IM53953
i-li, IM53961
i-li-a-ku-um, IM53961
i-l[*i*], IM53961
i-li-kum, IM53961
i-li, IM54559
i-li, IM54464
i-[*li-a*]*-kum*, IM54464
i-l[*i*], IM54464
i-li-kum, IM54464
i-[*li-a-kum*], IM54464
[*i-li*], IM54464

Comment: translated as to come up, with the figurative sense of to emerge, turn up.
emēdum (uš) v.; to lean against, reach, cling to, impose, put something on,

uš, IM55357 (see also *redûm*)

eperum n.; dust, earth, loose earth, debris, ore, mortar, territory, soil, area

e-pe$_2$*-ri*, IM54478
e-pe$_2$*-ri-ka*, IM54478
e-pe$_2$*-ru-ka*, IM54011
[*e-pe*$_2$*-ru-ka*], IM54011
e-pe$_2$*-ri-ka*, IM54011

Comment: translated as volume.
epēšum (du$_3$) v.; to do, act, make, build

ki$_3$.ta.zu.un.ne, IM55357
kid$_2$(?).zu$_2$.ne, IM52301
e-pe$_2$*-ši-ka*, IM54478
e-pe$_2$*-ši-ka*, IM53953
e-pe$_2$*-ši-*[*k*]*a*, IM54538
e-pe$_2$*-ši-ka*, IM53961
e-pe$_2$*-ši-ka*, IM53957
e-pe$_2$*-ši-ka*, IM54010
e-pe$_2$*-ši-ka*, IM53965
e-pe$_2$*-ši-ka*, IM54559
e-pe$_2$*-ši-ka*, IM54464
[*e-pe*$_2$*-ši-ka*], IM54011
e-pe$_2$*-ši-ka*, IM54011

Comment: translated as to do. For the expression za.e ak.ta.zu.ne, and the equivalent Akkadian *atta ina epēšika*, see the corresponding paragraph in Sect. 3.2.
eqlum (a.ša$_3$) n.; field, terrain, area, land, region

a.ša$_3$, IM55357
[a.š]a$_3$, IM55357
a.ša$_3$, IM52301
a.ša$_3$, IM53953
a.ša$_3$-*ka*, IM54010
a.ša$_3$, IM53965
a.ša$_3$-*ka*, IM53965
a.ša$_3$, IM54559

erbe num.; four

er-be$_2$*-e*, IM54010
er-be$_2$, IM54010 (see also *irbum*)

erbettum n.; a group or team of four

er-be$_2$*-tim*, IM52301

eṣēdum v.; to harvest

e-ṣi$_2$*-id*, IM54010

eṣēpum v.; to twine, to double

e-ṣi$_2$*-ma*, IM52301

ešer (fem. *ešeret*) num.; ten

e-še-re-[et], IM54538
e-še-re-et, IM54538

ezēbum (tag/k$_4$) v.; to leave, leave behind, abandon, desert, entrust, bequeath

ib$_2$.tag$_4$.a, IM55357 (see also *riāḫum*)

Comment: translated as to remain.
gamārum v.; to bring to conclusion, complete, bring to an end, destroy, use up, spend, settle, encompass, control, possess in full, finish, come to an end

li-ig!*-mu-ra-am*, IM54538
i-ga!*-ma-ru-ni-iš-ši*, IM54538
i-ta-ag-ma-ar, IM53957

gamrum (til) adj.; whole, complete, total, full, terminated

til, IM55357

ḫalāqum v.; to disappear, vanish, perish, become missing or lost, become fugitive, escape

ḫa-li-iq, IM52301
ḫa-al-qu$_2$, IM52301

ḫarāṣum v.; to cut down, cut off, deduct, cut deeply, determine, make clear

ḫu-ru-uṣ$_4$, IM52301
[*ḫu*]*-ru-iṣ*, IM53965
ḫu-ru-uṣ$_4$, IM54559
ḫu-ru-uṣ$_4$*-ma*, IM54464

Comment: The imperative *ḫuruṣ* in IM52301 is perhaps a parallel of the expected *ḫariṣ* (CAD Ḫ 92 b). The same applies to *ḫuriṣ* in IM53965.
ḫaṣābum v; to cut or break off

aḫ-ṣu$_2$*-ub*$_2$*-šu-ma*, IM53965
ta-aḫ-ṣu$_2$*-bu*, IM53965

ḫepûm v.; to smash, destroy, cut, crack, demolish, divide

ḫi-pe$_2$, IM55357
<*ḫi*>*-pe*$_2$, IM55357
ḫi-pe$_2$, IM52301
ḫi-pe$_2$*-ma*, IM52301

ḫi-pe$_2$*!(du*$^?$*)-ma*, IM52301
ḫi-pe$_2$*-e-ma*, IM53953
ḫi-pe$_2$*-ma*, IM54559
ḫi-pe$_2$, IM54011
ḫe-pi$_2$, IM54011
ḫi-pe$_2$*-e-ma*, IM54011
ḫe-pu-šu, IM54011
[*ḫi*]*-pe*$_2$*-ma*, IM54011

Comment: the AHw (I, 340) admits the imperative *ḫipe*. However, in the paradigm section of the GAG, *ḫipi* is the imperative (and *ḫepi* the stative). Gundlach and von Soden (1963) use the form *ḫe-pe*$_2$ for the imperative. In this work, I follow the option given by the AHw, and the verb is translated as "to halve".

ib$_2$.si$_8$ (Sumerian term) v.; "to make something equal" (see Terms for Square and Cube Roots in Sect. 3.2)

ib$_2$.si$_8$, IM55357
ib$_2$.si$_8$, IM54478

ib$_2$.si.e (Sumerian term) v.; "to make something equal" (see Terms for Square and Cube Roots in Sect. 3.2)

[i]b$_2$.si.e, IM53953
ib$_2$.si.e, IM53953
[ib$_2$].si, IM54010
ib$_2$.si.e, IM53965
ib$_2$.si.[e], IM54559
ib$_2$.si.e, IM54559

idûm v.; to know, be experienced or familiar with, be acquainted with

i-de-e!*-ma*!, IM53965

igi (Sumerian term) n.; reciprocal

igi, IM55357
igi, IM52301
i-gi, IM52301
i-gi, IM54478
igi, IM53953
i-gi, IM54538
i-gi, IM53961
i-gi, IM53957
i-gi, IM54010
[*i*]*-gi*, IM54010
[*i*]*-gi*, IM53965
i-gi, IM54464
i-gi-šu, IM54464
[*i*]*-gi*, IM54011
[*i-gi*], IM54011

Comment: kept untranslated *igi*. See Sect. 3.2.

igigubbûm n.; coefficient (math term)

i-gi-gu-ub-bi-im, IM52301
i-gi-gu-ub$^{?}$*-bi*, IM54538
i-gi-gu-bi-ka, IM54538
i-gi-gu-bi-ka, IM53961
i-gi-gu-[*ub-bi-ka*], IM54011

Comment: I assume that the stem of this term ends in the vowel *i*. Consequently, we should write *igigubbî*, the nominative of "my coefficient", with a circumflex.

ina prep.; in, inside, on, by, from

i-na, IM55357
i-na, IM52301
i-na, IM54478
i-na, IM53953
i-na, IM54538
i-na, IM53961
i-na, IM53957
i-na, IM54010
i-na, IM53965
i-na, IM54559
i-na, IM54464
i-na, IM54011

irbum n.; gifts, present, income, amount

ir-bi, IM54010 (see also *erbe*)

iškarum n.; work assignment, materials or supplies for workmen, finished products, staples or materials to be delivered

iš-ka-ar, IM53961
[*iš*]*-ka-ar*, IM54011
[*iš-ka-ar*]*-ka-ma*, IM54011

ištēn num.; one

iš-ten, IM52301
iš-ti-in, IM52301
iš-te-en, IM53961
iš-te-en, IM53965
[*iš*]*-te-en*, IM53965
iš-te-en, IM54559
[*i*]*š-te-en*, IM54559
iš-te-en, IM54464
iš-te-en, IM54011

kamārum v.; to heap up, pile up, accumulate, add

ku-mu-ur-ma, IM52301
ku-mu-ur, IM52301
ku-mu-ur, IM53953
i-ta-ak-ma-ar, IM53957
ku-mu-ur-ma, IM54464
ku-mu-ur, IM54011
[*ku-mu-ur*], IM54011

karûm n.; grain-heap, grain-store, pile of barley

ka-ru-um, IM52301

kaspum (kug.babbar) s; silver

kug.babbar, IM54464

kī (*kē*, *akī*, *akē*) interr.; how?

ki, IM54478
ki, IM54538
ki, IM53957
ki[?], IM54010
ki, IM54010
[*ki*], IM53965
[*ki*], IM54011

Comment: See also *maṣûm*.
kīam adv.; thus, in this manner

ki-a-am, IM52301
ki-a-am, IM54478
ki-ia, IM54478
ki-a-am, IM53953
ki-a-am, IM54538
ki-a-am, IM53961
[*ki*]-*a*-[*am*], IM53957
ki-a-[*am*], IM54010
ki-a-am, IM53965
ki-a-am, IM54559
ki-a-am, IM54464
[*ki-a-am*], IM54011
ki-a-am, IM54011

kippatum n.; loop, hoop, ring, circle, circumference, perimeter, totality

ki-pa-ti, IM52301

kubrum s; thickness, mass

[*ku-bu*]-*ur*, IM54011

kullum v.; to hold, maintain, keep, have, wear, rule

li-ki-il, IM52301
u$_2$*-ka-lu*, IM52301
li-ki-il, IM53965

Comment: for the expression *rēška likīl*, see Sect. 3.2.
kumurrûm n.; sum, total, accumulation, heaping up, piling up

ku-mu-ri, IM52301

kurrum (gur) n.; a measure of capacity, the amount of grain in one such unit

[gur]-*um*, IM53957

lā negative particle; no, not, without

la, IM52301

lapātum v.; to touch lightly, to touch in a symbolic way, touch, come in contact, attack, inscribe, write down, record

lu-pu-ut-ma, IM52301
lu-<pu>-ut-ma, IM52301
lu-pu-ut-ma, IM54478

leqûm v.; to take something in one's hand, take up an object (for a specific purpose), take objects or persons along, adopt, marry, accept, take over, take in

le-qe$_2$*-ma*, IM52301
el-qe$_2^?$*-a-ma*, IM53965

libbum n.; heart, entrails, womb, inner body, inside, interior

li-ib-bi-<šu?>, IM52301

libittum (sig$_4$) n.; brick, mud brick, brickwork, slab, block, cake (of material other than mud)

šeg$_{12}$, IM52301
li-bi-tu-um, IM54538
li-bi-ti-ka, IM54538

Comment: for value šeg$_{12}$ instead of sig$_4$ see ePSD (s.v. **šeg**). Notice, however, that šeg$_{12}$=sig$_4$.
maḫārum v.; to accept valuables, staples, persons; *šutamḫuru* to assume the same rank as someone else, to rival, to compete with someone, to confront

uš-ta-am-ḫi-ru, IM54478
uš-tam$^?$*-ḫi-ir*, IM54478

maḫīrum n.; market place, business transactions, exchange rate, purchase price

ma-ḫi-ir, IM54464

mala conj.; as much as, as many as, as many times as, everything that, everybody who

ma!-la!, IM52301
ma-la, IM54478

maṣûm v.; to correspond, comply, be equal to, be able to, be sufficient for, amount to

ma-ṣi$_2$, IM54478
ma-ṣi$_2$, IM54538
ma-ṣi$_2$, IM53957
ma$^?$*-ṣi*$^?_2$, IM54010
ma-ṣi$_2$, IM54010
ma-ṣi$_2$, IM53965
ma-[*ṣi*$_2$], IM54011

Comment: In the expression *kī maṣi*, "how many?", "how much?". See also *kī.*

maškanum, (ki.ud) n.; place of putting, threshing floor, site, settlement

ki.su$_7$-*im*, IM54538

Comment: according to ePSD, written ki.su$_7$ (s.v. **kisur** [LOCUS]). The word is not to be confused with ***maškānum*** n.; deposit, granary.

me'atu (me) num.; hundred

me, IM53957

mēlûm n.; height, altitude, high part (of a building or part thereof, of a person, an object), elevation, ascent, steps of a stair or ladder

me-li-um, IM53961
me-li-ka, IM53961

mi particle indicating direct speech

mi, IM53957

middatum n.; a certain measure of capacity, a certain measure of length, area and time, measuring rod, dimension, size, measure

[*mi-in-da-su*], IM53965

miḫrum (gaba.ri) n.; copy (of a written document), duplicate, replica, answer, reply, antiphony, fellow, equivalent, counterpart, front, correspondence, list, inventory

gaba, IM55357
me-ḫe-er-šu, IM52301

me-eḫ-ra-am, IM52301
me-eḫ-<<ša>>-ra-am, IM52301
me-eḫ-ra-am, IM53965
me-eḫ-ra-am, IM54559

mīnum (en.nam) interr.; what, why, what for, for what reason

mi-nu-um, IM55357
a.na(ba?).am$_3$, IM55357
mi-nu-um, IM52301
mi-nu, IM52301
mi-nam, IM54478
mi-nu-um, IM53953
mi-na-am, IM53953
mi-nu-um, IM53961
mi-na-am, IM54010
mi-na-am, IM53965
mi$^{?}$ [*mi*]*-nu-um*, IM54559
mi-na-am, IM54559
mi-nu-um, IM54011

mitḫartum n.; square, side of a square

mi-it-ḫa-ar-ta-ka, IM54478

Comment: translated as confrontation.

mitḫārum adj.; of equal size, amount, or degree, square, equal amount, corresponding, uniform, proportionate, equivocal, indecisive

mi-it-ḫa-ru-ti, IM52301

mūlûm s; height, hill, high ground, climb, ascent

mu-lu-um, IM54011
mu-li-im, IM54011
[*mu*]*-li-im*, IM54011

mušarum n.; surface measure of one square nindan, volume measure of one square nindan by one cubit

mu-ša-ar, IM54478
mu-ša-ri, IM54478
[*m*]*u*[*-ša-*]*ar*!, IM54538
mu-ša-ri, IM54538

Comment: translated with the logogram sar$_v$.

muttarrittum n.; perpendicular

mu-tar-ri-it-tum, IM55357

Comment: in mathematical contexts, better translated as descendant (Høyrup 2002, 229).

nadanum v; to give, offer, grant, share, pay, deliver, confer

i-na-di-na-ku-um, IM54464

nadûm v.; to throw, lay down, cast down, lay out, write, write down, make a drawing, impress a seal

i-di-ma, IM52301

naḫum (i$_{3}$.šaḫ) n.; lard, pig's fat

i$_{3}$.šaḫ, IM54464
na-ḫi-im, IM54464

napḫarum n.; sum, total, all, the whole, entirety, universe, totality

na!*-ap-ḫa-ar*, IM52301
na-ap-ḫa-<*ar*>, IM52301

nasāḫum (zi) v.; to remove, pull out, tear out parts of the body, take away, exterminate, deduct, subtract, excerpt a tablet, move on, displace oneself

ba.zi, IM55357
ta-na-sa-aḫ, IM52301
a-su-uḫ, IM54478
ta-su-ḫu, IM54464

našpakum n.; granary, silo, storehouse, capacity, storehouse vessel, cargo boat

na-aš-pa-kum, IM52301

našûm (il$_{2}$) v.; to lift, take up an object, raise, transport goods, etc., carry, bear, bring

il$_{2}$, IM55357
i-ši-ma, IM52301
i-ši-ma, IM54478
i-ši-ma, IM53953
i-ši-ma, IM54538
i-ši-i-ma, IM53961
[*i*]*-ši-i-ma*, IM53961
[*i*]*-ši-ma*, IM53957
i-ši-ma, IM54010
[*i*]*-ši-ma*, IM54010
i-ši-ma, IM53965
i-ši-ma, IM53965
i-ši-ma, IM54559
na-ši-a-ku, IM54464
i-ši-ma, IM54464

[*i-ši*]-*i-ma*, IM54464
[*i-ši-i-ma*], IM54464
i-ši-i-ma, IM54464
i-ši-[*ma*], IM54011
i-ši-ma, IM54011
[*i-ši-i-ma*], IM54011

Comment: the writing *i-ši-i-ma* indicates a lengthening of the final vowel of the imperative *iši* that may be attributed to the presence of the particle -*ma* (see GAG §15c). In our texts, it is translated as "bring" only in one occurrence, namely in the statement of IM54464.

nēpešum n.; activity, undertaking, doings, procedure, construction, ritual, instructions, tools, utensils, implements

ne-pe$_2$*-šum*, IM52301

nikkassum n.; a measure of length

ni-ka-as$_2$, IM54011

Commented: kept as *nikkassum* in the translation of IM54011.

parsiktum n.; a measure of capacity, mostly used for grain

3, for 3 parsiktum IM53957 (it is an uncertain reading; see the commentary of the problem)

paṭārum (duḫ, du$_8$) v.; to loosen, release, untie, detach, open, clear, dispel, dismantle

du$_8$.a, IM55357
duḫ.ḫa-ma, IM52301
pu-ṭu$_2$*-ur-ma*, IM52301
pu-ṭu$_2$*-ur*, IM52301
pu-ṭu$_2$*-ur-ma*, IM54478
pu-ṭu$_2$*-ur-ma*, IM53953
pu-ṭu$_2$*-ur*, IM54538
pu-ṭu$_2$*-ur-ma*, IM53961
pu-ṭ[*u*$_2$*-ur-ma*], IM53957
pu-ṭu$_2$*-ur-ma*, IM54010
pu-ṭu$_2$*-ur-ma*, IM53965
pu-ṭu$_2$*-ur-ma*, IM54464
pu-ṭ[*u*$_2$*-ur-ma*], IM54464
pu-ṭu$_2$*-ur-ma*, IM54011

pitiqtum n.; brickwork, mud wall, mud masonry

pi$_2$*-ti-iq-tum*, IM52301
pi$_2$*-ti-iq-tum*, IM53961
pi$_2$*-ti-iq-tum*, IM54011

pūtum (sag.ki) n.; forehead, brow, (short) side of a piece of immovable property, of a geometric figure, front, width

sag.ki, IM55357
sag (erasure), IM55357
sag.ki, IM52301
sag.ki, IM53953
pu-ta-am, IM53965
pu-tu, IM53965
pu-ti-ka, IM53965
pu-tu-u[*m*], IM53965
sag.ki, IM54559
[sag.ki], IM54559

qabûm v.; to say, tell, speak, promise, agree, pronounce, name, call

li-iq-bu-ni-kum-ma, IM52301

qanûm n.; reed, cane, arrow, tube, pipe, measuring rod, a measure of length, plot of land

qa-na, IM54010
qa-ni-ka, IM54010
qa-na-am, IM53965

qātum n.; hand, paw, handle

qa-ti-ia, IM52301

Comment: translated as hand, but also as possible metaphor for a computation device (see the commentary of IM52301 and Proust (2000)).

qûm (sila$_3$) n.; a measuring vessel of standard capacity, a measure of capacity

sila$_3$, IM53957
qa-a$^?$, IM54010
sila$_3$, IM54464

quppum n.; box, chest, basket, cage

qu$_2$-up-pi$_2$-im, IM52301

redûm (uš) v.; accompany, lead, drive, pursue, chase, continue, follow

uš, IM55357 (see also *emēdu*)

rešum n.; head, top, upper part, beginning, first instalment, original quantity, capital assets

re-eš$_{15}$-ka, IM52301
[*re-še$_{20}$*]*-um*, IM53957
re-še$_{20}$-e-ia, IM53957
re-še$_{20}$-e-im, IM53957
re-eš$_{15}$-ka, IM53965

Comment: translated as original quantity in IM53957.

riāḫum (tag$_4$) v.; to remain, be left over, be left behind, be spared, survive

ib$_2$.tag$_4$.a, IM55357 (see also *ezēbum*)

rupšum n.; width, breadth

ru-up-šu-um, IM53961
ru-up-šu-um, IM54011
ru-up-ša-am, IM54011
[*ru-up-ša*]-*am*, IM54011

saḫārum v.; to turn, turn around, turn back, stay around, persist

na-as$_2$-ḫi-ir, IM55357
na-as$_2$-ḫi-ir, IM53953
na-as$_2$-ḫi-ir, IM54010
na-as$_2$-ḫi-ir-ma, IM53965
na-as$_2$-ḫi-ir, IM54559
na-as$_2$-ḫi-ir-ma, IM54559
n[*a-as$_2$-ḫi-ir*]-*ma*, IM54559
na-as$_2$-ḫi-ir-ma, IM54464
na-as$_2$-ḫi-[*ir*]-*ma*, IM54464
na-as$_2$-ḫi-ir-ma, IM54011

Comment: the N-stem is translated here as to return.

sattakkum or ***santakkum*** (sag.du$_3$) n.; triangle, wedge, cuneiform wedge

sag.du$_3$, IM55357
<sag>.du$_3$, IM55357
sa$^{?}$-*ta*$^{?}$-[*ki*$^{?}$], IM52301
sa-*ta*$^{?}$-*ku-um*, IM53953

sūtum (ban$_2$) n.; a vessel, a measuring vessel of standard capacity and its volume, unity of capacity

ban$_2$, IM54464

ṣābum n.; people, workers, troops, army, population

ṣa$_2$-ba-am, IM54538
[*ṣa-bu-ka*], IM54011

ša det. pron.; of, that, which, that of, who(m)

ša, IM52301
ša, IM53953
[*ša*], IM53953
ša, IM54538
ša, IM53961

ša, IM53965
ša, IM54559
ša, IM54464

šakānum v.; to put, place, lay down, impose, establish, settle, set in place

šu-ku-un-ma, IM52301
ša-[ak-na-]at, IM54538
lu-uš-ku-un-ma, IM54538
šu-ku-un-ma, IM54538
šu-ku-un, IM53965

šalāšā num.; thirty

ša-la-ši!, IM53965

šālšum (3.kam) adj.; third, one third

3.kam, IM55357

šâlum **(*šaʾālum*)** v.; to ask, question, interrogate, inquire, investigate, ask permission

i-ša-al-ka, IM54478
i-ša-al, IM53953
i-ša-al-ka, IM54538
i-ša-al-ka, IM53961
[*i-ša-al-ka*$^?$], IM53957
[*i-ša-al-k*]*a*, IM54010
i-ša-al-[*ka*$^?$], IM53965
i-ša-al, IM54559
i-ša-al-ka, IM54464
i-ša-al-ka, IM54011

šamnum (i$_3$.giš) n.; oil, fat, grease, cream

i$_3$.giš, IM54464 (see also *ellum*)

šâmum (*šaʾāmu*) v.; to buy, purchase

ša-[*ma-am*], IM54464
ša-am, IM54464

šanûm (2.kam, min$_3$) adj.; second, next, other, another, second quality, second in rank, following

2.kam$_2$, IM55357
ša-ni-im, IM52301
min$_3$, IM52301
ša-ni-im, IM54478
ša$^?$*-ni-im*, IM54010

šapālum v.; to become low, go deep, reach the lowest point, bow low, suffer a loss; *šuppulum* (Š-stem) to lower, make lower, excavate, bring down from above, depress

u$_2$*-ša-pi*$_2$*-il-ma*, IM54478
u$_2$*-ša-pi*$_2$*-il*, IM54478

šapiltum s; lower part, inner part, second in rank, assistant, remainder

ša-pi$_2$*-il-tum*, IM54464

šaplûm (ki.ta) adj.; lower, lower-lying, of lower quality

ki.ta, IM55357
ša-ap-li-tim, IM52301
ša-ap-li-tum, IM52301
ša-ap-lu-um, IM53953
ša-ap-li-tim, IM53953
ša-ap-li-tum, IM53953

šārum n.; wind, direction, air, breath, emanation, emptiness

ša-ar, IM52301

Comment: the expression *šār erbettim* is translated in IM52301 as the four directions.

šeʾum (še) n.; barley, grain

še, IM53957
še$_{20}$*-e-ia*, IM53957
še$_{20}$*-e-im*, IM53957

Comment: for grain as a unit of measure, see *uṭṭatum*.

šiddum (uš) n.; long side, side, length, edge, bank

uš, IM55357
uš, IM52301
uš-*ia-ma*, IM52301
uš-*ia*, IM52301
uš-*ka*, IM52301
ši-di-ka, IM52301
uš, IM53953
[u]š, IM53953
ši-du-um, IM54538
ši-di-im, IM54538
ši-da-am, IM53965
ši-di, IM53965
ši-du-um, IM53965
uš, IM54559
ši-du-um, IM54011
[*ši-di-im*], IM54011

šina num.; <u>two</u>

ši-ta, IM53961
ši-ta, IM54011

šinipum (ša.na.bi) num.; <u>two thirds</u>

ši-ni-ip, IM52301
ši-ni-pe$_2$-tim, IM52301
ši-ni-ip-pe$_2$-at, IM52301
ši-ni-ip, IM53953
ša.na.bi, IM53953
ši-ni-ip, IM53957
ši-ni-pi$_2$-ia, IM53957
ši-ni-pi$_2$, IM53957
ša.na.bi, IM53957
ša.na.[bi], IM53957
ši-ni-ip, IM54559
ša.na.bi, IM54559
ši-ni-ip, IM54464

<u>Comment</u>: The logogram ša.na.bi appears as an alternative to the syllabic writing with a certain regularity in the analysed tablets. On the one hand, all but one syllabic writings bring a status constructus, either followed by a genitive or by a possessive suffix. The exception is *šinipêtim*, used in IM52301 in the expression "40 *šinipêtim* … *luput-ma*", meaning either "Record … 40 of the two thirds" or, as an apposition, "Record … 40, the two thirds". In the first case, *šinipêtim* performs the function of a genitive; in the second case, an accusative. As both are indicated by the same form in the plural, it is not possible to decide which one was the scribe's intention. On the other hand, the logogram ša.na.bi, as it occurs, in the analysed tablets, is never to be understood as a status constructus. Following the lesson of IM52301, it is therefore transcribed as *šinipêtim* in this work.

šinīšu (*šinīši*, *šinīša*) adv.; twice, a second time

ši-ni-i-ša, IM54010

šiqlum (gin$_2$) n.; <u>shekel</u> (a measure of weight, one sixtieth of a mina, circa eight grammes)

gin$_2$, IM54464

šittum n.; rest, <u>remainder</u>, remnant, balance

ši-ta-tum, IM52301
ši-ta-tum, IM53957

<u>Comment</u>: Used in singular and plural with approximately the same meanings.

šû pron.; he, that, this same

šu-u$_2$-ma, IM54478
šu-u$_2$-ma, IM53953

šu-u$_2$*-ma*, IM54538
šu-u$_2$-[*ma*], IM53961
[*šu-u*$_2$*-ma*], IM53957
šu-u$_2$*-ma*, IM54010
[*šu-u*$_2$*-ma*], IM53965
šu-u$_2$*-ma*, IM54559
šu-u$_2$*-ma*, IM54464
šu-u$_2$*-ma*, IM54011
šu-u$_2$-[*ma*], IM54011

šumma indecl.; if, whether, either-or

šum-ma, IM52301
šum-ma, IM54478
šum-ma, IM53953
šum-ma, IM54538
šum-ma, IM53961
šum-ma, IM53957
šum-ma, IM54010
šum-ma, IM53965
šum-ma, IM54559
šum-ma, IM54464
[*šum-ma*], IM54011
šum-ma, IM54011

šuplum n.; depth, deepness, minimum latitude

šu-pu-ul-ka, IM54478

šūši num.; sixty

šu!-[*ši*], IM53965

takiltum n.; a mathematical term

ta-ki-il-ta-ka, IM52301

Comment: kept untranslated *takiltum*. See Sect. 3.2 for explanation.

tallum (dal) n.; a type of container or vessel

tal-<li>, IM53957
dal, IM53957

târum v.; to return, to return to a previous position or status, become again, repeat, retreat

tu-ur-ma, IM52301

u conj.; and, but, also

u$_3$, IM55357
u$_3$, IM52301

u$_3$, IM54478
u$_3$, IM53953
u$_3$, IM53957
u$_3$, IM54010
u$_3$!, IM53965
u$_3$, IM54464
u$_3$, IM54011

ul negative particle; not

u$_2$*-ul*, IM53965

ūmakkal adv.; 1 (single) day, all day long, for the length of 1 day

u$_2$*-ma-ka-al*, IM54538

Comment: according to von Soden (1952, 51), *umakal*. Translated here as in 1 day.

ūmakkalûm adj.; (enough, required, sufficient) for 1 day, assigned for 1 day

u$_2$*-ma-ka-li-a-am*, IM54538
u$_2$*-ma-ka-lu-tu-un*, IM54538
u$_2$*-ma-ka-lu-tum*, IM54011

Comment: According to von Soden (1952, 51), *umakalûm* or *umakalium* "zu einem Tag gehörig".

umma particle introducing direct and indirect speech

um-ma, IM54478
um-ma, IM53953
[*um-ma*], IM54538
um-ma, IM53961
[*um-ma*], IM53957
um-ma, IM54010
[*um-ma*], IM53965
um-ma, IM54559
um-ma, IM54464
um-ma, IM54011

Comment: translated here as saying.

uṭṭatum n.; grain, barley, kernel, grain (a unit of measure)

še, IM54464

wabālum v.; to bring, carry, carry off, fetch, transport

ta-ba-al-ma, IM53957

warkûm (according to ePSD, eğir$_6$=MURGU$_2$)) adj.; future, later in time, second, lower on rank, back, rear

murgu$_2$, IM55357 (see also *būdum*)

waṣābum (taḫ) v.; to enlarge, add, increase in size or number

ṣi$_2$*-ib-ma*, IM52301
ṣi$_2$*-[ib]-ma*, IM52301
taḫ$_2$*-ma*, IM52301
tu-iṣ-bu, IM52301
ṣi$_2$*-im-ma*, IM52301
u$_2$*-ṣi*$_2$*-im-ma*, IM53957
ṣi$_2$*-ib*, IM53965
u$_2$*-ṣi*$_2$*-[ib]*, IM54559
taḫ$_2$*-im-ma*, IM54559
ṣi$_2$*-ib*, IM54559
ṣi$_2$*-ma*, IM54464

watārum v.; to exceed in number or size, surpass in importance or quality, (be) come outsize

e-te-er, IM52301
i-te-ru, IM52301
i-te-er, IM52301
e-te-ru, IM52301
wa-ta-ar, IM54559
wa-at-ri-im, IM54464
wa-at-r[i]-im, IM54464

Comment: in mathematical texts, with preposition *eli*, translated as to go beyond.

zūzum n.; half, half unit, half-shekel, half-sila

zu-uz$_4^?$, IM54478

References

Dictionaries, Sign List and Grammar

AHw: von Soden's Akkadisches Handwörterbuch (vol. I, 1985, Zweite Auflage, vol. II, 1972, vol. III, 1981). Otto Harrassowitz, Wiesbaden

CAD: The Assyrian Dictionary of the Oriental Institute of the University of Chicago, commonly quoted as the Chicago Assyrian Dictionary (1956–2010). 21 volumes

GAG: von Soden, W. Grundriss der Akkadischen Grammatik. 3, ergänzte Auflage, unter Mitarbeit von Werner R. Mayer. Editrice Pontificio Istituto Biblico (1995)

ePSD: The Pennsylvania Sumerian Dictionary. http://psd.museum.upenn.edu/epsd/index.html

MesZL: Borger, R. Mesopotamisches Zeichenlexikon. Zweite, revidierte und aktualisierte Auflage. Ugarit, Münster (2010)

Other

Notes and statistics. Sumer **II**, 13–18 (1946)

al-Fouadi, A.-H.: Lenticular Exercise School Texts. Texts in the Iraq Museum, vol. X. Republic of Iraq—Ministry of Culture and Arts, The State Organization of Antiquities, Baghdad (1979)

al-Rawi, F.N.H., Roaf, M.: Ten Old Babylonian mathematical problems from Tell Haddad, Himrin. Sumer **XLIII**, 175–218 (1984)

Attinger, P.: À propos de AK "faire" (I). Zeitschrift für Assyriologie **95**, 46–64 (2005)

Attinger, P.: Racines carrées et racines cubiques. Zeitschrift für Assyriologie **98**, 12–19 (2008)

Baqir, T.: Excavations at Tell Harmal. Sumer **II**, 22–30 (1946). Plus 6 planches

Baqir, T.: Excavations at Harmal. Sumer **IV**, 137–139 (1948)

Baqir, T.: Date-formulae and date-lists. Sumer **V**, 34–86 (1949a)

Baqir, T.: Supplement to the date-formulae from Harmal. Sumer **V**, 136–143 (1949b)

Baqir, T.: An important mathematical problem text from Tell Harmal (on a Euclidean theorem). Sumer **VI**, 39–54 (1950a). Plus 2 planches (IM55357)

Baqir, T.: Another important mathematical text from Tell Harmal. Sumer **VI**, 130–148 (1950b). Plus 3 planches (IM52301)

C. Gonçalves, *Mathematical Tablets from Tell Harmal*, Sources and Studies in the History of Mathematics and Physical Sciences,
DOI 10.1007/978-3-319-22524-1

Baqir, T.: Some more mathematical texts from Tell Harmal. Sumer **VII**, 28–45 (1951). Plus 5 planches (IM54478, IM53953, IM54538, IM53961, IM53957, IM54010, IM53965, IM54559, IM54464, IM54011)

Baqir, T.: Tell Harmal. General Antiquities Directorate, Baghdad (1959)

Baqir, T.: Foreword. Sumer **XVII**, 1–12 (1961) (esp. pages 1–4)

Baqir, T.: Foreword. Sumer **XVIII**, 5–14 (1962). Plus three planches

Bouzon, E.: Uma Coleção de Direito Babilônico Pré-Hammurabiano. Editora Vozes, Petrópolis, RJ (2001)

Bruins, E.M.: Comments on the mathematical tablets of Tell Harmal. Sumer **VII**, 179–185 (1951) (includes a letter to the editor and a "note added during correction")

Bruins, E.M.: Revision of the mathematical texts from Tell Harmal. Sumer **IX**, 241–253 (1953a)

Bruins, E.M.: Three geometrical problems. Sumer **IX**, 255–259 (1953b) (not from Tell Harmal. IM43996, IM31248)

Bruins, E.M.: Some mathematical texts. Sumer **X**, 55–61 (1954) (from Tell Harmal: IM52001, IM54346, IM54216, IM52548, IM55111, IM55292, IM52879, IM54486, IM54472, IM52672; from Ischali: IM31210)

Bruins, E.M.: On the system of Babylonian geometry. Sumer **XI**, 44–49 (1955)

Bruins, E.M., Rutten, M.: Textes mathématiques de Suse (Mémoires de la Mission Archéologique en Iran 34). Geuthner, Paris (1961)

Cavigneaux, A.: A scholar's library in Me-Turan? With an edition of the tablet H 72 (Textes de Tell Haddad VII). In: Abusch, T., van der Toorn, K. (eds.) Mesopotamian Magic. Textual, Historical and Interpretive Perspectives, pp. 215–273. Styx, Groningen (1999)

Charpin, D., Ziegler, N.: Mari et le Proche-Orient à l'époque amorrite. Essai d'histoire politique. Florilegium marianum V. SEPOA, Paris (2003)

Cherry, J.F.: Peer polity interaction. In: Renfrew, C., Bahn, P. (eds.) Archaeology. The Key Concepts, pp. 147–150. Routledge, London (2005)

Civil, M.: Ancient Mesopotamian lexicography. In: Sasson, J.M. (ed.) Civilizations of the Ancient Near East, pp. 2305–2314. Charles Scribner, New York (1995)

Drenckhahn, F.: A geometrical contribution to the study of the mathematical problem-text from Tell Harmal (IM55357) in the Iraq Museum, Baghdad. Sumer **VII**, 22–27 (1951)

Drenckhahn, F.: Ein geometrischer Beitrag zu dem mathematischen Problem-Text von Tell Harmal IM 55357 des Iraq Museums in Baghdad. Zeitschrift für Assyriologie und Vorderasiatische Archäologie, Neue Folge, Band **16**, 151–162 (1952a)

Drenckhahn, F.: Letter to the editor. Sumer **VIII**, 234–235 (1952b)

Duby, G.: L'histoire continue. Odile Jacob, Paris (1991)

Friberg, J.: Bricks and mud in metro-mathematical cuneiform texts. In: Høyrup, J., Damerow, P. (eds.) Changing Views on Ancient Near Eastern Mathematics, 61-154. Dietrich Reimer Verlag, Berlin (2001)

Friberg, J.: Mathematik. In: Edzard, D.O., et al. (eds.) Reallexikon der Assyriologie und Vorderasatischen Archäologie, vol. 7, pp. 531–585. Walter de Gruyter, Berlin (1987–1990)

Friberg, J.: Mathematics at Ur in the Old Babylonian period. Revue d'Assyriologie **94**, 97–188 (2000)

Friberg, J.: On the alleged counting with sexagesimal place value numbers in mathematical cuneiform texts from the third millennium BC. Cuneiform Digit. Libr. J. 2 (2005). http://cdli.ucla.edu/pubs/cdlj/2005/cdlj2005_002.html

Friberg, J.: A Remarkable Collection of Babylonian Mathematical Texts. Springer, New York (2007a)

Friberg, J.: Amazing Traces of a Babylonian Origin in Greek Mathematics. World Scientific, Singapore (2007b)

George, A.R.: In search of the é.dub.ba.a: the ancient Mesopotamian school in literature and reality (2001). http://eprints.soas.ac.uk/1618/1/GeorgeEdubbaa.pdf

Goetze, A.: The laws of Eshnunna. Sumer **IV**, 63–102 (1948)

Goetze, A.: A mathematical compendium from Tell Harmal. Sumer **VII**, 126–155 (1951). Plus 8 planches

Goetze, A.: Fifty Old Babylonian letters from Harmal. Sumer **XIV**, 3–78 (1958). Plus planches

Gonçalves, C.H.B.: Tabletes Matemáticos Cuneiformes e a Questão da Materialidade (Cuneiform mathematical tablets and the issue of materiality). In: Anais do 12o Seminário Nacional de História da Ciência e da Tecnologia e 7o Congresso Latino-Americano de História da Ciência e da Tecnologia. Sociedade Brasileira de História da Ciência, Salvador, Bahia (2010)

Gonçalves, C.H.B.: A Álgebra no Século XVII. In: Muhana, A., et al. (eds.) Retórica, pp. 31–44. Annablume, Instituto de Estudos Brasileiros, São Paulo (2012)

Gonçalves, C.H.B.: Quantification and computation in the mathematical texts of the Old Babylonian Diyala. In: Chemla, K., Keller, A., Proust, C. (eds.) Cultures of Computation and Quantification (forthcoming)

Gundlach, K.-B., von Soden, W.: Einige altbabylonische Texte zur Lösung 'quadratischer Gleichungen'. Abhandlungen aus dem mathematischen Seminar der Universität Hamburg **26**, 248–263 (1963)

Heron-Heiberg, J.L.: Heronis Definitiones cum variis collectionibus, Heronis quae feruntur Geometrica. In: Schmidt, W., et al. (eds.) Heronis Alexandrini opera quae supersunt omnia. 5 vols. and supplement, pp. 1899–1914. Teubner, Leipzig (1912)

Høyrup, J.: IM52301. Conformal translation and commentary. Typed manuscript. 8 pages (no date)

Høyrup, J.: Algebra and naive geometry. An investigation of some basic aspects of Old Babylonian mathematical thought II. Altorientalische Forschungen **17**, 262–354 (1990)

Høyrup, J.: Changing trends in the historiography of Mesopotamian mathematics: an insider's view. History of Science **xxxiv**, 1–32 (1996)

Høyrup, J.: Lengths, widths, surfaces. A portrait of Old-Babylonian algebra. Springer, New York (2002)

Høyrup, J.: State, 'justice', scribal culture and mathematics in ancient Mesopotamia. Sartoniana **22**, 13–45 (2009)

Høyrup, J.: How to transfer the conceptual structure of Old Babylonian mathematics: solutions and inherent problems. With an Italian parallel. In: Imhausen, A., Pommerening, T. (eds.) Writings of Early Scholars in the Ancient Near East, Egypt, Rome and Greece. Translating Ancient Scientific Texts, pp. 385–417. Walter de Gruyter, Berlin (2010)

Høyrup, J.: A note about the notion of exp10(log10(modulo 1))(x) Concise observations of a former teacher of engineering students on the use of the slide rule. In: Contribution au Séminaire SAW: Histoire des mathématiques, histoire des pratiques économiques et financières. Séance du 6 janvier 2012: "Usage de la position—pratiques mathématiques, pratiques comptables" (2012)

Hussein, L.M.: Tell Harmal. Die Texte aus dem Hauptverwaltungsgebäude "Serai". Inaugural-Dissertation zur Erlangung der Doktorwürde dem Fachbereich Fremdsprachliche Philologien der Philipps-Universität Marburg (2009)

Hussein, L.M., Miglus, P.A.: Tell Harmal—Die Frühjahrskampagne 1997. Baghdader Mitteilungen **29**, 35–46 (1998)

Hussein, L.M., Miglus, P.A.: Tell Harmal—Die Herbstkampagne 1998. Baghdader Mitteilungen **30**, 101–112 (1999)

Isma'el, K.S., Robson, E.: Arithmetical tablets from Iraqi excavations in the Diyala. In: Baker, H.D., Robson, E., Zólyomi, G. (eds.) Your Praise Is Sweet. A Memorial Volume for Jeremy Black from Students, Colleagues and Friends, pp. 151–164. British Institute for the Study of Iraq, London (2010)

Kouwenberg, N.J.C. The Akkadian verb and its Semitic background. Eisenbrauns, Winona Lake, IN (2010)

Lion, B., Robson, E.: Quelques textes scolaires paléo-babyloniens rédigés par des femmes. J. Cuneiform Stud. **57**, 37–54 (2005)

MCT: Neugebauer and Sachs (1945)

Mieroop, M. Van De. A History of the Ancient Near East ca. 3000 - 323 BC. Blackwell, Oxford (2007)

Miglus, P.A.: Šaduppûm B. In: Streck, M.P., et al. (eds.) Reallexikon der Assyriologie und Vorderasatischen Archäologie, vol. 11, 5/6, pp. 491–494 (2007); vol. 11, 7/8, p. 495 (2008). Walter de Gruyter, Berlin (2007, 2008)

MKT: Neugebauer (1935–1937)

Muroi, K.: Mathematical term *Takīltum* and completing the square in Babylonian mathematics. Historia Scientiarum **12**, 254–263 (2003)

Nemet-Nejat, K.R.: Cuneiform Mathematical Texts as a Reflection of Everyday Life in Mesopotamia. American Oriental Series, vol. 75. American Oriental Society, New Haven, CT (1993)

Netz, R.: The Shaping of Deduction in Greek Mathematics: A Study in Cognitive History. Cambridge University Press, Cambridge (1999)

Neugebauer, O.: Mathematische Keilschrifttexte, I–III. Springer, Berlin (1935–1937)

Neugebauer, O., Sachs, A.: Mathematical Cuneiform Texts. American Oriental Series, vol. 29. American Oriental Society, New Haven, CT (1945)

Powell, M.A.: The antecedents of Old Babylonian place notation and the early history of Babylonian mathematics. Historia Mathematica **3**, 417–439 (1976)

Powell, M.A.: On the verb AK in Sumerian. In: Dandamayev, M.A., et al. (eds.) Societies and Languages of the Ancient Near East: Studies in Honour of I. M. Diakonoff, pp. 314–319. Aris and Phillips, Warminster (1982)

Powell, M.A.: Maße und Gewichte. In: Edzard, D.O., et al. (eds.) Reallexikon der Assyriologie und Vorderasatischen Archäologie, vol. 7, pp. 457–516. Walter de Gruyter, Berlin (1987–1990)

Proust, C.: La multiplication babylonienne: la part non écrite du calcul. Revue d'histoire des mathématiques **6**, 293–303 (2000)

Proust, C.: Tablettes mathématiques de Nippur. Varia Anatolica, vol. XVIII. Institut Français Georges Dumézil, De Boccard, Paris (2007)

Proust, C.: Du calcul flottant en Mésopotamie. La Gazette des Mathématiciens **138**, 23–48 (2013)

Renfrew, C.: Introduction: peer polity interaction and socio-political change. In: Renfrew, C., Cherry, J.F. (eds.) Peer Polity Interaction and Socio-political Change, pp. 1–18. Cambridge University Press, Cambridge (1986)

Ritter, J.: Reading Strasbourg 368: a thrice-told tale. In: Chemla, K. (ed.) History of Science, History of Text. Boston Studies in the History and Philosophy of Science, vol. 238, pp. 177–200. Springer, Dordrecht (2005)

Robson, E.: Mesopotamian Mathematics, 2100–1600 BC. Technical Constants in Bureaucracy and Education. Oxford Edition of Cuneiform Texts, vol. XIV. Clarendon Press, Oxford (1999)

Robson, E.: Mathematical cuneiform tablets in Philadelphia. Part 1: problems and calculations. SCIAMVS **1**, 11–48 (2000)

Robson, E.: Neither Sherlock Holmes nor Babylon: a reassessment of Plimpton 322. Historia Mathematica **28**, 167–206 (2001)

Robson, E.: Mesopotamian mathematics. In: Katz, V. (ed.) The Mathematics of Egypt, Mesopotamia, China, India and Islam: A Sourcebook, pp. 58–186. Princeton University Press, Princeton, NJ (2007)

Robson, E.: Mathematics in Ancient Iraq: A Social History. Princeton University Press, Princeton, NJ (2008)

Robson, E.: Mathematics education in an Old Babylonian scribal school. In: Robson, E., Stedall, J. (eds.) The Oxford Handbook of the History of Mathematics, pp. 199–227. Oxford University Press, New York (2009)

Sjöberg, Å.: Der Examenstext A. Zeitschrift für Assyriologie **64**, 137–176 (1974)

Sjöberg, Å.: The Old Babylonian Eduba. In: Lieberman, S.J. (ed.) Sumerological Studies in Honor of Thorkild Jacobsen on His Seventieth Birthday. Assyriological Studies 20, pp. 159–179. University of Chicago Press, Chicago, IL (1976)

Sulaiman, A.: Harvest documents and loan contracts from the Babylonian period. Sumer **XXXIV**, 130–138 (1978)
Thureau-Dangin, F.: Esquisse d'une histoire du système sexagésimal. Geuthner, Paris (1932)
Thureau-Dangin, F.: Textes Mathématiques Babyloniens. Brill, Leiden (1938)
Tinney, S.: Texts, tablets and teaching. Scribal education in Nippur and Ur. Expedition **40**(2), 40–50 (1998)
TMB: Thureau-Dangin (1938)
TMS: Bruins and Rutten (1961)
Veldhuis, N.: Elementary education at Nippur: the lists of trees and wooden objects. Doctoral thesis, University of Groningen (1997). http://socrates.berkeley.edu/~veldhuis/EEN/EEN.html
von Soden, W.: Zu den mathematischen Aufgabentexten vom Tell Harmal. Sumer **VIII**, 49–56 (1952)

Printed in the United States
By Bookmasters